MAPPING COMMUNITY HEALTH

APPLYING GIS

MAPPING

COMMUNITY

HEALTH

GIS FOR HEALTH & HUMAN SERVICES

Edited by

Christopher Thomas
Shannon Valdizon
Matt Artz

Esri Press
REDLANDS | CALIFORNIA

CONTENTS

INTRODUCTION

H EALTH AND HUMAN SERVICES OFFICIALS WORK DAILY TO improve health outcomes, increase access to health care, and build healthier communities. They recognize the impact on their organizations when they use location to understand coverage gaps and risks to individuals.

Preparing for health emergencies and responding to human crises such as the opioid epidemic, homelessness, food insecurity, and health and racial inequities have placed a renewed emphasis on geographic approaches to many complicated health challenges. Building communities that improve health outcomes and increase accessibility to health care is critical. The coronavirus disease 2019 (COVID-19) pandemic illustrated and raised the importance of geography in public health preparedness and response. At the same time, the dynamic nature of this kind of crisis makes it difficult to develop and deploy timely solutions.

Health and human services professionals can benefit from insights that the power of place brings to their work. Spatial data and geographic information system (GIS) technology can underscore coverage gaps in the populations they hope to reach, create opportunities to improve workflows, and help them plan for and respond to unforeseen events. Applied GIS leads to improved health outcomes, better access to health care, and healthier communities.

Stories and strategies

This book presents a collection of real-life stories that illustrate how organizations are using GIS to help humans in crisis, address public health readiness, confront the challenges of health equity, ensure access to health care, and support strategic planning. The book concludes with a section about getting started with GIS, which provides ideas, strategies, tools, and suggested actions that organizations can take to build location intelligence into decision-making and operational workflows. The stories and strategies aim to help the reader understand how to use GIS to gain a geographic perspective and integrate spatial reasoning into health and human services operations. The book presents location intelligence as another crucial layer of knowledge that managers and practitioners can add to their existing experience and expertise, offering a unique, geographic perspective they can incorporate into daily operations and planning.

If location intelligence isn't currently part of an organization's decision-making processes or considered in daily operational activities, or if it isn't used to improve patient satisfaction, then managers can use this book to start developing skills in those areas. Developing these skills does not require GIS expertise, nor does it require managers to disregard all their experience and knowledge. Spatial reasoning adds another way to think about solving problems in a real-world context.

HOW TO USE THIS BOOK

THIS BOOK IS DESIGNED AS A GUIDE TO HELP YOU TAKE first steps with GIS to address issues that are important to you right now. It will help you apply location intelligence to decisions and operational processes for solving common problems and creating a more collaborative environment in your organization. You can use this book to identify where maps, spatial analysis, and GIS apps might be helpful in your work and then, as next steps, learn more about those resources.

Learn about additional GIS resources for health and human services by visiting the web page for this book:

go.esri.com/mch-resources

PART 1

HUMANS IN CRISIS

SOLVING PROBLEMS FOR HUMANS IN CRISIS HAS BEEN A recent rallying point for health and human services professionals. GIS provides a framework for crisis response—to organize data geographically, receive real-time updates, communicate information to stakeholders, deploy tactics, allocate resources, and keep decision-makers and the community informed.

Prepare for health emergencies

Health emergency preparedness involves extensive planning for a wide range of contingencies, including escalating operational support and shifting strategies that focus resources where and when they are needed most. GIS helps users quickly understand the common operational picture and provides tools to support the framework for crisis response.

Track disease

Monitoring diseases and conditions that can cause serious public health threats is critical to safeguard and build healthy communities. Tracking the source and spread of a disease or a pandemic such as COVID-19 can be enhanced by adding location and demographics. Location data tells a more complete story that can help communicate patterns and provide insight into mitigating needs.

Reduce homelessness

The homeless crisis requires a modern, integrated, multidisciplinary approach. To address this social problem, we must consider housing people who experience homelessness, reducing disease in encampments, connecting people to services, understanding the needs of communities of color, and identifying root causes. Because location data connects these disparate factors, GIS provides a unique view into building effective policy and maximizing resources.

Combat vector-borne disease

Protecting the public from vector-borne disease requires proactively reducing, monitoring, and controlling vector populations. GIS provides a foundation from which to support the end-to-end workflows, from identifying source points and dispatching crews and materials to performing analysis, informing management, and keeping the public informed.

Address the opioid crisis

The opioid epidemic constitutes a state of emergency in many places. GIS is a proven technology that benefits human services, health organizations, and law enforcement. Mapping and analyzing their data together shows a crisis that requires local governments to look for insights, pinpoint the sources of the problem, and deliver effective response plans and allocation of resources.

GIS in action

This section will look at real-life stories about how health and human services organizations use GIS to count the homeless, battle infectious disease, understand the opioid crisis, and more.

DATA AND SMART MAPS ELEVATE HUMAN SERVICES

Snohomish County, Washington State

I N MID-JANUARY 2020, A MAN IN SNOHOMISH COUNTY, Washington, was dubbed "Patient Zero" of COVID-19 in the United States. The diagnosis set off a countywide response effort.

Officials scrambled to understand what was then a new threat and focused on vulnerable populations, including the area's unsheltered population. As time went on and the true nature of the crisis became apparent, the county's solution helped reach hundreds of homeless people in need of shelter, essential resources, and medical services.

The work started in March 2020 when Alessandra Durham, a senior policy analyst with the Snohomish County Executive's Office working in the Snohomish County Emergency Coordination Center, created the SnoCo Agencies for Engagement (SAFE) team to conduct outreach efforts. The SAFE team is composed of physicians, community paramedics, social workers, and law enforcement officers. The team spends time visiting unsheltered people to assess needs and connect to needed services.

To ensure success, the team took a targeted, data-driven approach supported by advanced technology. In the field, the SAFE team began using mobile apps powered by location intelligence from the county's GIS to identify exactly where specific services were needed. The information was shared in real time with other teams in every appropriate department for planning and decision making.

"A lot of times, we'll know what to do to help either mitigate or resolve issues, but having accessible real-time data has been really helpful in effectively deploying very limited resources," said Durham.

"Utilizing GIS has really helped us to better inform and target where we deploy evidence-based practices."

Becoming a data-driven county

In 2018, the Snohomish County Human Services Department received a federal Data-Driven Justice Initiative grant. Launched in 2016 under President Obama, the initiative was designed to break the cycle of over-incarceration of vulnerable populations by aligning justice, health, and human services systems around real-time data. The goal was to identify frequently incarcerated people and effective ways to divert them away from the justice system and into community-based services and treatment providers.

Nate Marti, planning and evaluation division manager of the Human Services Department, was one of the key architects in developing a data-driven approach to community outreach. He says the grant opened the door for Human Services to begin using location intelligence for the first time.

"The county has had experts using GIS products for years, but mainly within other departments," Marti said. "Prior to 2018, human services really never touched the geospatial analysis that we're using today."

One of the first things Marti did was launch a pilot program to develop a baseline understanding of the county's unsheltered population. They developed a field application that outreach teams could use to identify the locations and demographics of homeless encampments.

All of the data collected was entered directly into the mobile surveying app, ArcGIS® Survey123, and displayed visually on an interactive web map that could be shared with other teams within the Outreach Coalition, a collaboration across multiple agencies and organizations that coordinates outreach to individuals and families.

For instance, the outreach teams collected information about potential hazards—such as needles or animals—at each camp, so support staff knew what to expect when visiting a location.

In another example, outreach teams tried to determine if any veterans lived in the camps. "Then we could deploy veteran outreach services to that encampment so they could receive the appropriate services," Marti explained.

Any department could search for encampments that included the specific populations it served, such as the elderly, or it could filter for that information on the map to identify the encampments the department wanted to focus on.

Keeping unsheltered populations safe during COVID-19

When the pandemic hit Snohomish County in early 2020, staff within the Emergency Coordination Center knew they must act quickly to ensure the safety of one of the most vulnerable populations. The county was concerned about the spread of COVID-19 among individuals experiencing homelessness, because oftentimes, they lived close together in congregate shelters or encampments without access to sanitation resources and showers, Marti said.

To properly deploy the right social services from the SAFE team, the first task was to understand the locations and demographics of the region's current homeless population.

The outreach team included medical staff to assess COVID-19 symptoms. A housing navigator connected people to housing resources. Embedded social workers linked people to behavioral health services, law enforcement, the fire department, and emergency services.

The team sought a baseline understanding similar to what they gained from a pilot program two years earlier—but with an additional layer of data collected about how the population was being affected by the virus.

Marti said the previous experience proved key to getting the program off the ground quickly. Given the many different stakeholders, it was an accomplishment that would have taken much longer without a collaborative technology like GIS.

"Because of our previous experience using Survey123 to collect homeless encampment data, we were able to develop a survey and deploy the application within five days," Marti said.

Over two separate weeklong outreach periods in the spring of 2020, the team contacted more than 400 unsheltered individuals. The team members were surprised to find very few people displaying COVID-like symptoms. Instead, the primary way they had been affected by the virus was in their access to essential resources like food, water, showers, restrooms, shelter, and medical treatment.

The SnoCo Agencies for Engagement (SAFE) team includes doctors, paramedics, social workers, and law enforcement officers who have gone out in the field to reach the county's most vulnerable residents. Their work included spending three evenings visiting meal distribution sites in Everett, Washington, to assess people with COVID-19 symptoms and provide hygiene kits and face coverings.

As parks, restaurants, and offices closed, the unsheltered population was left with nowhere to turn for basic amenities.

The outreach team used that insight to deploy crews into the field to support people who needed assistance with behavioral health or housing, for instance. The county's Emergency Coordination Center also used that information to create hundreds of individual hygiene kits, which the SAFE team dispersed to people experiencing homelessness.

Marti says the app's ease of use and data-capturing capabilities in real time proved essential during the early days of the crisis, especially among fieldworkers who had little experience using GIS.

"The team was able to easily input the information on a mobile app as it was being gathered," he said. "And it's real time, so there's no lag in the data. Especially early on with COVID-19, we needed answers fairly quickly. This just provided a really easy mechanism to gather the information and then share it."

Durham says that for those with a nontechnical background, one of the biggest attractions of GIS is its ability to democratize data. "The tools make information digestible and accessible to people like myself and to the county executive."

Understanding a crisis

The COVID-19 pandemic is not the first time the county used GIS to support human services. The Snohomish County Opioid Response Multi-Agency Coordination (MAC) Group also used GIS to help show how opioid abuse affects residents.

The MAC Group started by collecting data from a number of different agencies, including fire and emergency medical services (EMS), law enforcement, and social services. When any of those agencies responds to an opioid-related call, that data goes into a GIS-powered dashboard where it's displayed on a smart map. If an emergency medical technician uses a drug called Narcan to counter-act an opioid overdose, for instance, that data is automatically sent to the dashboard, so the team can see when the drug was deployed, demographic data about the patient, and where the event took place.

Using a GIS dashboard to analyze and display data gave the MAC Group important insights about the extent of the opioid crisis in Snohomish County. The dashboard has also helped county offi-cials effectively share information with the public, thanks to the uni-versal appeal of maps. People began to see that no neighborhood was immune from opioid abuse when they saw a map of places where emergency responders attended opioid-related calls.

Durham said the biggest surprise for the public has been seeing exactly whom the epidemic is affecting. "It's not only the geographic locations of opioid-related events but also the age range that this impacts. We see very, very young people and people into their 70s. Displaying the data visually helps the general public better under-stand the depth and breadth of the issue—that it's not other people. It's about them and their families as well."

The use of GIS has helped county become more proactive in its response. For instance, county leaders can identify locations with a higher incidence of use and deploy intervention services to help pre-vent overdoses. With limited resources, targeted outreach is essential.

Marti believes that, given its wide breadth of capabilities, GIS has already become critical to the county. "It's been two years since Human Services began using GIS, but I think we're still just scratch-ing the surface with the technology that's available."

A version of this story by Christopher Thomas titled "COVID-19: Data and Smart Maps Are Elevating Human Services in Snohomish County" originally appeared in the *Esri Blog* on September 29, 2020.

GIS INSPIRES A NEW WAY FOR THE POINT-IN-TIME COUNT

Placer and Nevada Counties, California

P LACER AND NEVADA COUNTIES EXTEND OVER RURAL LAND east of San Francisco, and like other counties, they are federally mandated to conduct a Point-in-Time Count (PTC) to survey the homeless population in their communities. The PTC is critical to secure resources and understand at-risk populations. The counties are part of the Placer-Nevada Counties Continuum of Care (CoC), committed to the goal of ending homelessness by connecting services and resources more effectively.

There must be a better way

In years past, Placer County depended on paper surveys to conduct the PTC. To prepare for the count, the county printed and provided all the surveys and tally count sheets. Volunteers, including county staff and nonprofit service providers, go into the field and conduct the surveys. Once the surveys were returned, county staff manually entered the data into a spreadsheet, a process that took about 50 hours. The work included sorting through papers and trying to interpret the handwriting of the volunteers.

With no accurate way to capture their location, volunteers would estimate by street cross-sections. Back in the office with the survey results, staff had to locate the street cross-sections and manually plot the points on a map. Results were displayed in a heat map that only gave estimated locations of people experiencing homelessness.

The county also faced the challenge of understanding and compiling the data into reports in a timely manner. The effort to input data and ensure the data was clean without duplicates or missing elements took many days.

Making the count, count

In October 2018, as the counties prepared for the 2019 PTC, Sue Compton, homeless management information system administrator at Placer County, was introduced to GIS mobile data collection tools. As she and her team readied for the next count, Compton recognized the value GIS could bring to increase efficiency in the PTC. Compton presented a strategy to the CoC board to use GIS as the foundation for the PTC to receive resources to support people experiencing homelessness. Compton turned to the Placer County GIS team to explore the possibilities of enhancing data collection for volunteers in the field and presenting the collected information efficiently and accurately through reports for decision-makers.

ArcGIS Survey123 allowed them to send volunteers in the field with a mobile survey, which provided an accurate location, cut down concerns around data security from carrying around paper surveys, and allow Placer County to create maps and run reports at a quicker rate. Using Survey123 relieved the pressure of leaving room for error when the count would be taken.

This proposed mobile strategy also answered staff concerns about effectively carrying out the PTC in Nevada County. Nevada County stretches over more rural land and is harder to reach than Placer County and relies on its staff and volunteers to conduct the count. Training an additional 50 volunteers would have required much more work and even a walk-through on how to collect information. Ensuring that the additional help was all on the same page was critical to executing an efficient PTC strategy. The application's simplistic user interface sped up Placer County's training on how to collect information and optimized staff's time.

On the day of the PTC, volunteers used their smartphones to easily and quickly collect information while in the field. While volunteers collected data, the information channeled to the office in real time, and the team conducted quality assurance (QA) on the collected surveys. Incorporating GIS into the strategy allowed the team to accurately collect data in a fraction of the time, meet federal US Department of Housing and Urban Development (HUD) requirements, report to stakeholders more quickly, increase understanding of the crisis, and present opportunities to intervene with policy.

A volunteer fills out a survey of questions that are required in the PTC.

Raising the bar for future counts

By embracing a geospatial strategy, the Placer County team and volunteers worked efficiently to survey and count the number of people experiencing homelessness. "With GIS, we are able to improve the efficiency of our homeless count and more accurately identify areas where we can target services and resources to help reduce homelessness in our community," said Compton. Using Survey123 on their mobile devices allowed volunteers flexibility to collect federally required HUD information to better understand the unique needs of each community and create a performance dashboard.

After analyzing the data and visualizing where people experiencing homelessness live, the Placer-Nevada Counties CoC increased its insight into the homelessness crisis in their communities. The counties plan to enhance the use of GIS for future Point in-Time Counts to see how to continue to foster their geospatial strategy into their ecosystem and use real-time data.

A version of this story by Nadine Hernandez titled "How GIS Inspired a New Way for the Point-in-Time Count" originally appeared on Esri.com in 2020.

INFECTIOUS MOSQUITOES POSE INCREASING HEALTH THREAT

Volusia County, Florida

MOSQUITOES REPRESENT A GRAVE THREAT TO HUMAN LIFE. With more than 3,500 species and their ability to adapt quickly to new environments, mosquitoes have tremendous potential to spread disease-causing pathogens around the world.

In the past 15 years, we've seen examples of mosquito-borne diseases that have caused illness, disability, and death at a pandemic scale. Tens of thousands of people in the United States have suffered serious cases of West Nile virus and Zika virus, to name a few. Zika virus also has the terrible potential consequence of causing microcephaly, a type of cranial deformity, in the babies of women infected during their pregnancy. In Florida, where the mosquito threat persists year-round, Zika transmission has increased by 60 percent since 2004 after larvae from the Asian tiger mosquito larvae arrived inside imported tires filled with pools of water bound for Texas in 1985. Wet conditions caused by hurricanes and other rain events exacerbate the problem.

Explosions of mosquito vector populations have raised the attention of the United States and other national governments to investigate the crisis and respond to it with equal measures of brains and brawn. The "brains," in this case, consist of a location technology platform that helps organize, coordinate, and guide the prevention, surveillance, and control activities of an integrated vector-control plan. The "brawn" is in the applied tactics that public health and vector control agencies use to protect human populations around the world.

GIS is core equipment in the fight

With mosquitoes carrying an assortment of tropical fevers and uncommon viruses, there's great interest in adopting and applying modern technology in new and effective ways. GIS is a key example of such a modern platform.

Geographic thinking is a natural fit for vector-borne disease surveillance and control. The geographic approach to such daunting challenges is increasing because GIS technology provides a framework for mapping and modeling that can be easily applied to any vector control operation. A location perspective streamlines and coordinates the workflow and approach from surveillance to response.

The foundation of vector control

A proper reconnaissance must start with two fundamental questions:

1. Where are the mosquitoes that we should be concerned about?

2. How do they behave?

Without a scientific truth source—datasets and databases—tactical response is impossible. Using GIS to address the mosquito threat begins with surveillance activities. Mosquito surveillance involves predicting where the potentially infected insects will be so that counter-assaults can be strategically targeted.

The distribution of mosquitoes is a delicate fly, restricted by specific biological tolerances. determined by environmental variables such as elevation, climate, precipitation, and land cover. GIS facilitates prediction of a vector's preferred environments by visualizing the environmental datasets and databases. Imagine trying to consider all of these factors in a spreadsheet. It would be impossible to detect geographic patterns of habitat suitability.

Volusia County, Florida, deploys an integrated mosquito management plan using many tools for its mosquito fight, including helicopters to scout for larvae and apply ultra-low-volume insecticide spray over broad areas when surveillance indicates that infestations are widespread.

Engaging the human sensor network

Nowadays, given the increasing use of smart devices and an interest in greater engagement in issues that impact us all, crowdsourcing supplements surveillance data in the vector fight. After all, we experience the mosquito bites and see the potential mosquito breeding grounds during our daily commutes and evening walks. But traditional mechanisms don't necessarily motivate us to report our data. We may get caught up in endless phone loops or struggle to find the right e-mail contact to express our concern. When we manage to make a report, the precise geographic context may be missing.

The mobile phone provides a fast and easy way to photograph and share our eyewitness accounts. More and more counties, such as Atlantic County, New Jersey, use location intelligence so that citizens can provide the most critical intelligence in the fight through an incident reporter app called Mosquito Service Request. It's quick work to configure GIS-based incident reporter apps that link to city

Staff from the Volusia County Public Works Department access maps and work orders from the field, record surveillance findings, and document their mitigation activities.

databases and alert staff to incidents in real time. This capability can result in a far more strategic response than sending an e-mail that must go through an extended chain of command before anything is done.

Gathering real-time surveillance information from passive activity is another tool. Social media posts can be collected and mapped, using specific tags to enhance agency knowledge of potential issues that may not otherwise be reported.

Measured response calms safety concerns

Vector control strategies always must address concerns about the pesticides used on established vector populations. The first GIS project by the author of this article, Este Geraghty, looked into the effect of aerial pyrethrin pesticide spraying on exposed human populations in Sacramento, California, in 2005. Vector control leadership knew that the treatment was effective against mosquitoes, but it was

Volusia County uses spraying trucks for more prescriptive applications when surveillance indicates that nuisance populations of mosquitoes are concentrated and easily reached via roadways.

important to ensure that populations did not suffer harm from pesticide exposure.

Maps oriented and focused that study in a way that raw spreadsheets couldn't. For example, the flight paths and spraying swaths from planes that spread the treatment were digitally captured and used to develop precise exposure models. Levels of exposure were compared to the addresses of people who visited emergency rooms for any potential complaint related to a pesticide exposure. In the end, the study found that the aerial application of pyrethrin pesticides, using the ultra-low volume (ULV) method of spraying, was not associated with any increase in emergency room visits.

Public health pesticide use, when applied to the right areas at the right times, is critical to preventing serious illness from mosquito-borne disease. Using GIS to make those decisions, while also continually monitoring for any potential adverse effects provides people with the dual benefit of reduced risk and safe practices.

Rapid response reduces risk of disease

Vector control agencies empowered by a location intelligence platform can quickly understand prediction and surveillance information to reduce vector transmission. Such responders can streamline their regular workflows with location intelligence.

Here are just a few parts of the workflow that can be made more efficient:

- Determine where to set traps and which types of traps to set.

- Associate laboratory results with trap locations on a map.

- Review service requests spatially to create appropriate treatment zones.

- Identify exclusion zones, such as schools, on the map.

- Assign jobs to field staff based on proximity to the job.

- Help field technicians navigate from their current location to their assignment location.

- Allow supervisors to track the status of field assignments in real-time.

A location-driven approach to vector-borne disease surveillance amplifies counterresponse and gives us strategic edge in the war. With record rainfall and flooding in Texas and Florida, it's hard not to think about new mosquito hatcheries causing harm if left unreported. Cities that give the power of surveillance to residents gain important allies in the fight.

A version of this story by Este Geraghty originally appeared in *WhereNext* on September 15, 2017.

RESEARCHERS UNLOCK OPIOID OVERDOSE PATTERNS USING GEOSPATIAL ANALYSIS

University of Texas Health Science Center at Houston/
Houston Emergency Opioid Engagement System

O PIOID OVERDOSES HAVE CLAIMED MORE THAN 400,000 lives since 1999, according to the Centers for Disease Control and Prevention. The epidemic has accelerated to the point where Americans are more likely to die from an overdose than a car accident or shooting. These startling statistics prompted the president to declare opioid use disorder a national public health emergency.

James Langabeer, a cognitive scientist and health services researcher, is tackling this epidemic at both a national and local level. Langabeer is a professor of biomedical informatics, emergency medicine, and public health at The University of Texas Health Science Center at Houston (UTHealth), where His research focuses on the determinants of death and treatment success. He also leads the Houston Emergency Opioid Engagement System (HEROES) program, where the Houston Fire Department, Houston Police Department, Houston Recovery Center, and Memorial Hermann Hospital-Texas Medical Center collaborate to provide a comprehensive system of care for managing opioid use disorder.

This conversation discusses the use of GIS in both Langabeer's research and outreach, touching on proven patterns and practices.

This interview has been edited and condensed.

Q: How do maps and spatial analytics aid you in understanding trends in opioid use disorder?
A: We can run statistical and analytical models and come up with fancy equations, but they don't often put data in terms that people

can readily understand. Visual analysis and geospatial analysis specifically have proven more useful in describing problems and pinpointing where we should place our efforts.

Q: Do you use GIS in your daily practice?
A: Yes, we map everything. Geospatial models are very important to us on a daily basis. We map where people are dying from drug overdoses using data from the local medical examiner's office. We also map and analyze where people are overdosing based on data from the fire department. We look at national data from the CDC. We use all available data to identify priority zones to target.

We also use the data to detect changes over time, comparing this month or this quarter with data from the past. This helps us understand progressions of drugs, movement of dealers, and the disease. We don't necessarily understand why it's happening, but the data and the analysis helps us to understand where it is, and to try to dig deeper for the triggers.

Q: Have you uncovered any patterns to understand the spread of opioid overdoses across the country?
A: We have developed a statistical model with county-level data on social, health, and economic determinants. We looked at the relationship between where people are dying and the opioid mortality rate. Our goal is to figure out which of these factors are most associated with death. In some areas, the determining factor could be the economy.

We have also looked at factors that determine what keeps people from staying in recovery. Homelessness is a factor as are medical issues, including hepatitis C or HIV.

We're working on interpreting our data and digging deeper to understand it.

Q: What's different about how the HEROES program tackles opioid use disorder?

A: The HEROES program takes a long-term approach that links medical treatment, behavioral treatment, and peer recovery support. We have fostered a community-wide collaboration, getting police, fire, and hospital emergency departments to talk with traditional outpatient medical treatment providers, psychiatrists, and family medicine practitioners. Traditionally, these groups do not communicate regularly.

Normally, treatment is one-on-one, with little collaboration regarding an individual with opioid use disorder. We're trying to link everyone in the public safety and health care community around people who have this problem.

Q: Have you achieved results with your data-driven actions?

A: We have had success rapidly identifying individuals who have overdosed or are in frail condition with a high potential for an overdose. We work to identify them, pinpoint their location, and get people mobilized to initiate treatment proactively rather than waiting for them to show up for treatment.

We send a paramedic and a peer recovery support specialist, someone who has the disease but has been sober many years, to do assertive outreach. They knock on the doors of people who have overdosed in the past and try to recruit them into treatment at no cost to them. A number of people come into treatment that have never committed to treatment before.

We know we're saving lives because we've studied the number of overdoses and the number of deaths occurring locally, and we're able to see the impact.

Q: What's your approach to treatment?

A: We really need to think about treating this disease with a multi-factorial approach with a lot of things in place. Our treatment begins with medical treatment. We prefer people to be on buprenorphine rather than go cold turkey. You have to be in some kind of behavioral treatment as well. Every participant in our program goes through addiction counseling and they have access to licensed psychologists. The third component of our approach is to surround participants with positive peers that have made it through treatment.

We really think you have to have all three of these things in order to adapt the brain for a long-term recovery. It's very difficult to break the chains of this disease.

Q: In your recent paper in the *Journal of Addiction Medicine*, you look at disparities between opioid overdose deaths and treatment capacity across the US. Do you think meaningful change will require policy changes?

A: We see a lot of policy changes that are necessary in terms of creating new capacity, as well as how providers are incentivized to treat this disease. In a full third of the country there is a dearth of treatment capacity. It's just not existent.

Even where the capacity exists, the wait times can be very long. Another real problem is that the treatment centers may only take people with cash or with a certain insurance. People with opioid use disorder have a brain disorder that causes a series of bad choices. Many lose their job, their health insurance, their family and friends, and everything else. Policy change is necessary to create flexible and affordable treatment capacity.

To reduce the harmful effects of the epidemic, where people are using substances and dying from them, we need better public health policies. Programs being considered include safe injection sites, needle exchange programs, making Narcan overdose treatment widely available, and other harm reduction policies. All can have an impact.

A version of this story by Este Geraghty originally appeared in the *Esri Blog* on June 27, 2019.

PART 2

PUBLIC HEALTH PREPAREDNESS

G EOGRAPHIC INFORMATION SERVES AS THE COMMON
denominator for any public health preparedness solution
because location data is essential to every phase of planning,
response, and recovery. GIS scales to events ranging from inclem-
ent weather conditions to pandemics. It helps state and local governments determine the extent of a public health crisis and predict
the path forward for allocating resources and monitoring processes and outcomes. Maps and apps enable public health officials
to coordinate efforts with other agencies and external stakeholders.
The public health preparedness community can make major progress by embracing GIS data, models, communication and engagement hubs, and locationcentric applications.

Resource allocation

A high-impact response and recovery plan requires the right
resources in the right place at the right time. GIS can forecast
surge in demand and optimize resource allocation to help place
resources precisely where they are most needed. Mapping and
analysis can expose gaps in capacity, identify the needs of at-risk
populations, help effectively distribute supplies, and inform per-
sonnel allocation. Location intelligence supports the distribution
of resources where people and needs intersect.

Decision support

Using location intelligence effectively in a health emergency preparedness plan requires the ability to gather and analyze data quickly and draw on experience from past events. Data-driven decision-making is about using accurate data to answer fundamental questions at a moment's notice. GIS is a crucial technology because so many questions hinge on location data.

Communication and collaboration

No matter how large or small the health response, situational awareness is critical. Every health crisis requires cross-sector support, including public safety, human services, and health care. These stakeholders need a common operational picture of what is happening and where so that they can communicate and collaborate effectively. GIS provides a range of communication and collaboration tools to support these efforts.

Civic engagement

Civic engagement can take many forms before, during, and after a natural disaster or health emergency. Location provides context for an emergency response that offers a more intuitive approach for public engagement and risk communication. People become more engaged when they understand how a crisis affects where they live, work, learn, and play. GIS supports community education through storytelling with maps, public notifications, and citizen input.

GIS in action

This section will look at a real-life story about the use of GIS to support contact tracing during the COVID-19 pandemic.

GIS-BASED CONTACT TRACING INITIATIVE SETS US PRECEDENT

Allentown, Bethlehem, York City, and Wilkes-Barre, Pennsylvania

WHEN VICKY KISTLER, DIRECTOR OF HEALTH FOR Allentown, Pennsylvania, and Matt Leibert, the city's chief information officer, traveled together to an Esri conference in Philadelphia in 2003, they imagined that one day they might have an advanced GIS platform that could help them visualize and track the spread of disease in real time. Little did they know that 17 years later, they would adopt this capability to confront a global pandemic.

Today, Allentown has taken the lead on a four-city, GIS-based community contact tracing initiative alongside Bethlehem, York City, and Wilkes-Barre that is poised to set the foundation for one of the COVID-19 pandemic.

The limitations of an app-based model

When the pandemic first hit, many local governments turned to app-based solutions for contact tracing and exposure notification to try and keep up with the spread of the virus. Unfortunately, cities often found themselves in situations where app-based models failed to garner momentum, resulting in adoption rates below those required to be effective. So state and local governments relied largely on traditional manual processes to build contact tracing workforces.

As the pandemic continued, transmission advanced to community spread, where those who get infected cannot pinpoint how they contracted the virus. This makes identifying and tracking contacts much more difficult. In turn, it's necessary to adopt a location-focused approach.

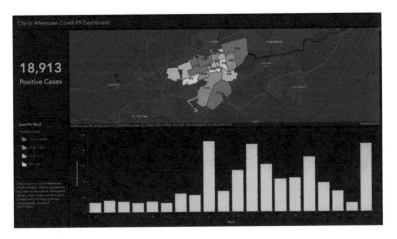

The City of Allentown's dashboard for the COVID-19 pandemic protects people's privacy while providing a clear visualization of what's going on.

But the problem is far more complicated than downloading apps and building up the workforce, according to Matt Leger, policy research analyst for the Innovations in Government Program at Harvard Kennedy School. Leger, who also serves as director of strategy at CONTRACE Public Health Corps, which provides COVID-19 contact tracing workforce and advisory solutions to public and private sectors, said tracing programs face operational constraints aggravated by often-outdated back end IT infrastructure that slows responses.

Combining person-to-person tracing with upgraded IT infrastructure and GIS tools, however, can help overcome these limitations.

Layers of benefits to using GIS

Allentown has taken this combined approach. As Kistler and Leibert explained, using community contact tracing tools from Esri produced many layers of benefits.

First, these tools gave Allentown an advantage because it already used ArcGIS. Allentown invested in IT solutions, including broadening its GIS capacity, several years ago—even in a period of resource scarcity—hoping that the investment would increase productivity. This foresight continues to produce benefits for routine and extraordinary service demands. Although the GIS capacity differs among the four cities participating in the community contact tracing initiative, their access to ArcGIS tools provided the foundation for standardization and collaboration.

These tools also helped Allentown replace cumbersome, paper-based processes with digital technology and partially-automated workflows. Leibert and his team analyzed the workflow of COVID-19 case investigation and contact tracing processes to identify duplication. They quickly realized that the city needed far more advanced capabilities to do effective contact tracing.

Officials identified nine essential manual steps between receiving data from the National Electronic Disease Surveillance System (NEDSS) and sending the data back to the Pennsylvania Department of Health—a labor-intensive effort filled with risks of human error. This case investigation process did not include additional steps needed for contact tracing.

Joseph Yashur, community health specialist for the City of Bethlehem, agreed that successful contract tracing required new capabilities. He noted that Esri® tools such as ArcGIS Survey123 performed tasks, including quality assurance on the data, much more efficiently.

Myriad of advantages for the community

The four Pennsylvania cities participating in this initiative found that automated and spatially oriented workflows that connect case investigation and contact tracing produce substantial advantages as community contact tracing accelerates. Examples of the many manual steps that have been automated to improve efficiency include verifying addresses and notifying contact tracing staff when there is a rapid reassignment of cases.

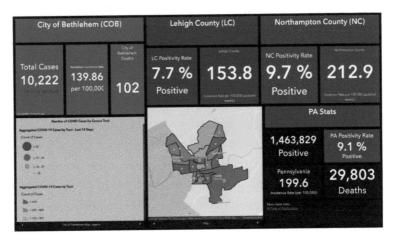

Officials from the City of Bethlehem agree that cities and counties benefit from sharing data and borrowing technology ideas from each other.

But as Joe McMahon, former managing director for Allentown's mayor, Ray O'Connell, said, "It's not just a matter of efficiency; it's a matter of accuracy." The original contact tracing platform provided to the City of Allentown by the state's department of health was not equipped for the level of detail needed to do contact tracing at scale, which would have affected accuracy.

"You know the old saying—if you enter it once, it's less likely that there will be transferred errors throughout multiple systems," McMahon added.

Yet the spatial nature of contact tracing improves the ability of city leaders to inform the public and enact strategic, place-based interventions. So accuracy is key.

The GIS-based contact tracing tools that Allentown, Bethlehem, York City, and Wilkes-Barre have put in place help protect people's privacy while providing dashboards and map narratives of COVID-19 cases to visualize what is going on. These capabilities, in turn, help improve community safety.

Geographic analysis can also assist by connecting the locations of the disease, affected individuals, and nearby available resources. Allentown intends to integrate an automated notification system to help communicate information to high-risk neighborhoods and to people who have tested negative or been exposed to someone who has tested positive.

Looking ahead to future public health uses

Cloud-based platforms allow public health departments to easily share innovations and discoveries. Kistler hopes that as other cities in her collaborative begin using the technology, "they may have ideas we haven't thought of that could enhance our capabilities even further."

"We're all borrowing from one another. It's information sharing that we're looking at doing," said Kristen Wenrich, Bethlehem's director of health. And that's something she takes pride in. "I'm anticipating that we'll hold regular meetings, since we're all speaking in the same language of GIS," she added.

A foundation based on GIS allows multiple agencies to gain even broader insight into public health. Bethlehem and Allentown public health leaders hope that the layered data will help cities focus more attention on the social determinants of health. Officials are considering how to apply GIS technology to trace sexually transmitted diseases (STDs), tuberculosis, whooping cough, and other illnesses. And the four cities are optimistic that, by crossing health data with

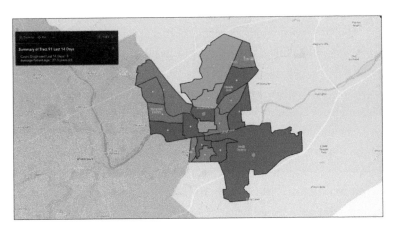

Geographic analysis can help city officials make connections among the locations of COVID-19 cases, affected individuals, and available resources.

geographic data, they can join together to apply for research grants and respond to the causes of health disparities.

According to Wenrich, Bethlehem plans to have officials—when they become more comfortable with the power of spatial analysis—use these new capabilities for other disease prevention activities.

"When you see mapped clusters, you can also see potential action that can be taken," she said.

For example, these capabilities helped Bethlehem improve outreach efforts to Spanish-speaking communities after the public health department saw the disproportionate impact of COVID-19 in certain Hispanic neighborhoods. Wenrich is contemplating applying this technology more broadly—for food inspections, public health nuisance complaints, and highway safety initiatives, for example.

The officials also hope that the increased efficiencies they've experienced will allow the state's public health department to make better decisions about other important tasks, such as mass flu shot and COVID-19 vaccine distribution, when the latter becomes available.

Often, according to McMahon, the public doesn't understand the need for technology investments. But, he said, "the transition to this platform [is] an investment that will have massive returns for not just public health but [also] city government overall."

A version of this story by Stephen Goldsmith titled "GIS-Based Contact Tracing Initiative in Pennsylvania Sets US Precedent" originally appeared in the Fall 2020 issue of *ArcNews*.

PART 3

HEALTH EQUITY

NEVER BEFORE HAS THERE BEEN SUCH A FOCUS ON equity as it relates to health—especially to ensure that health and human services organizations address the social determinants of health and gaps in social and racial equities. These inequities involve poverty, age, living conditions, and education, and it's important to understand the connection between where things are happening and where service gaps exist. A location-based strategy can help create an effective approach to collecting data; performing analysis; allocating resources where they're needed most; communicating with decision-makers; and, ultimately, achieving health equity.

Address population health management

To deliver programs and services that align with needs requires understanding a community's makeup today and what it will be in the future. Looking at demographic, socioeconomic, and lifestyle data builds a more complete community picture, helping to explain health disparities, analyze public health trends, and prepare for future needs.

Generate community health assessments

Community health assessments must do more than fulfill a legal requirement; they should be actionable. They should give insight into who needs the most help and where to allocate resources. Incorporating GIS into community health assessments adds the ability to run spatial analysis to see hot spots, determine where resources will have the greatest impact, understand how to lead health interventions, and decide how to better serve patients and prepare for their needs.

Citizen engagement

Building a community requires two-way communication. GIS allows effortless information sharing in an easy-to-understand format to give residents an accurate snapshot of their community's health, nearby services, and authoritative information. GIS-based maps of available resources—parks, fresh-food markets, services, health insurers, and more—inform the community and, more importantly, elicit citizen participation and feedback for community-based key initiatives.

Performance monitoring

Successfully using GIS to address health inequities involves using readily available applications and dashboards that improve situational awareness, allow users to shift to iterative policy making, and address issues in real time. These apps allow organizations to set baselines, monitor progress in at-risk areas, and offer performance transparency within the community.

GIS in action

This section will look at real-life stories about how health and human services organizations use GIS to map trees, urban parks, food and other essentials, and more.

HOW A MAP OF TREES HELPED CITY LEADERS TACKLE SOCIAL INEQUITIES

Austin, Texas

IN AUSTIN, TEXAS, LIKE MANY PLACES, THE NUMBERS OF trees in neighborhoods mark a divide of race and income.

For Austin, the correlation can be seen in tree canopy maps that city staff have overlaid with demographics and other data using GIS. In west Austin—the area west of Interstate 35—tree canopy covers 78 percent of the land. In east Austin, tree canopy covers only 22 percent.

"It's really interesting that Interstate 35 is also a dividing line for ecoregions," said Alan Halter, a senior GIS analyst with the City of Austin. "If you go west, you get into the Hill Country, with a lot more tree cover, but east Austin hasn't historically supported as many trees. And when you look at who lives where throughout Austin's history, communities of color have resided in east Austin."

Inequity in numbers of trees

In 1928, Austin's development plan relegated the city's Black communities to a district east of present-day Interstate 35. Restrictive zoning and real estate tactics had the impact of making nearly impossible for those residents to move but placed fewer restrictions on white residents to purchase homes in Austin's heavier-canopied parts of town.

The 1950s locked in environmental injustices, when the planning commission zoned all east Austin property as "industrial," affecting nearby residents with the area's lower air quality, higher temperatures because of a lack of tree cover, and other health-related issues.

This common pattern is found throughout the world—the prevailing winds, blowing west to east, bring pollution to the eastern parts of town.

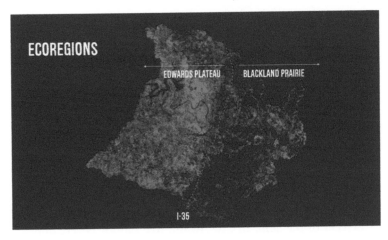

The ecoregion divide shows a sharp disparity in tree cover, noted in green.

When Halter first mapped Austin's trees, he was focused on the city council's urban forest plan.

"I created the first Community Tree Priority map back in 2015, and it was really tree-planting oriented—to figure out where to plant trees," Halter said. "At that time, equity was a consideration but wasn't really a main focus. We were mostly looking at where tree canopy existed and didn't exist, with the idea to increase shade across town and get trees where they're not currently located."

As Halter added layers of data to the map, he saw the relationship between socially vulnerable neighborhoods and areas with minimal trees. In the wake of Black Lives Matter protests, equity became a strong focus of the Austin Community Climate Plan, so the map needed to change.

"We released an update in 2020 with equity as the driving force," Halter said. "We're now looking at tree planting to achieve positive outcomes for people, such as improved public health; reduced heat-island effects; and, of course, addressing climate change, because it's related to everything."

The same line that denotes ecoregions also marks a racial divide in Austin.

To understand the impacts of having fewer trees, the city participated in an urban heat island mapping project coordinated by the federal government's National Integrated Heat Health Information System, a collaboration between the National Oceanic and Atmospheric Administration (NOAA) and the Centers for Disease Control and Prevention (CDC).

"Volunteers drove different routes in their cars with fancy devices poking out of their windows that recorded temperatures every few seconds," Halter said. "Using GIS, we could extrapolate temperature readings on a larger scale to see what heat looks like around town and compare it to tree canopy."

The result was an interactive web map showing that morning temperatures were higher in dense urban areas close to the city center than in other areas. Results exposed how concrete structures that absorb solar heat in the day and radiate it at night can be seven degrees hotter than outlying areas in the day and five degrees hotter at night. Strategies to mitigate the effects of urban heat islands include white roofs; more crosswalks so that people don't have to walk as far; more bus shelters; and, of course, more trees.

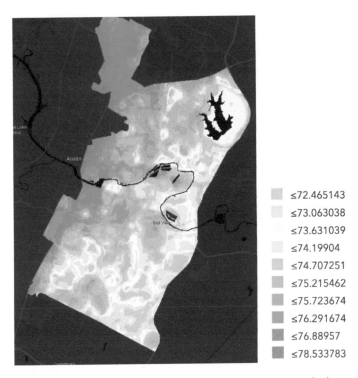

	≤72.465143
	≤73.063038
	≤73.631039
	≤74.19904
	≤74.707251
	≤75.215462
	≤75.723674
	≤76.291674
	≤76.88957
	≤78.533783

Interactive web map showing that morning temperatures were higher in dense urban areas close to the city center (red) than in other areas of Austin (blue).

The effect of trees on the lives of residents

Planting more trees in an underserved area starts a positive chain reaction: more trees mean more canopy; more canopy means more shade; more shade means less heat; less heat means lower energy bills and more outdoor activity. Therefore, more trees result in improved health and quality of life.

Trees create fresh air while also cleaning some pollutants. The greenery is appealing, which draws people outside where they can move around and be more social. A recent study even found that street trees are present where fewer people take medications to deal with depression. Trees also actively cool areas in a process that's similar to perspiration.

"The scientific term is evapotranspiration," Halter says. "I noticed it on a superhot day in July at midday, and suddenly these trees started to cry or sweat, as if it was raining. The trees are taking up water from the ground, then it goes up to the leaves, and then the tree rains on itself and the water goes back into the soil. It's kind of a breathing, liquid-to-gas process."

This analogy is fitting, since forests are often called the lungs of the earth, but most people don't experience the process so directly.

"Trees cool the environment—you can actually feel it," Halter said.

This measurable benefit is often referred to as an ecosystem service.

Trees also help protect areas from increasingly severe storms—especially important in a place like Austin, which experiences frequent cycles of drought and flooding. Tree roots draw in rainwater and keep the soil from washing away. The leafy limbs slow heavy raindrops before they hit the ground, so the soil is less prone to erosion.

Selecting the right trees

In 2014, the US Forest Service (USFS) inventoried trees in Austin to help understand tree canopy in detail and assess the carbon sequestration capacity of trees.

USFS analysts determined that Austin had 33.8 million trees, which stored about 1.9 million tons of carbon dioxide. Researchers found that every year, the trees remove about 92,000 tons of carbon and 1,253 tons of air pollutants, and reduce residential energy costs by $18.9 million.

The inventory included a species review, finding that the most common trees are Ashe juniper, cedar elm, live oak, sugarberry, and Texas persimmon.

Ashe juniper, Halter said, is "the number one cause of tree allergens in Austin—but it's also the tree species with the greatest air quality impact because it's an evergreen species and Ashe juniper trees act as year-round air filters. Ashe juniper also captures the most stormwater runoff and sequesters the most carbon in our urban forest. It's a weird dichotomy—the tree that's disliked the most is also the tree that's helping us the most."

Like all urban-forestry specialists, Halter and his colleagues in Austin have a difficult management task dealing with weather and infestations. Oak wilt, a fungal disease that is spread by beetles and gets into the root system, can spread from tree to tree underground. The emerald ash borer is a nonnative pest that hasn't hit yet, but Austin is getting ready because once it's present, it typically wipes out the entire ash tree population.

"We don't yet know the impact climate change will have on pests like borers or the fungus causing oak wilt, but we can expect things to get worse with a rise in temperature, which increases the stress on trees," Halter said. "We have tree doctors that go out and give shots to the trees for things like oak wilt, but it's really tough for trees to respond and survive if they're not getting enough water to begin with."

Trees, equity, and social justice

According to Austin's climate plan, the city "recognizes historical and structural disparities and a need for alleviation of these wrongs by critically transforming its institutions and creating a culture of equity." The city gives away free trees every year, but it has only recently examined—from an equity perspective—where those trees have been planted.

"We have to look at who has the means to drive across town on a Saturday to pick up trees," Halter said. So trees are now offered at giveaway events located in previously underserved neighborhoods. "With canopy mapping, we can assess it and show people, through the data, what that looks like. And the community tree priority map—with scoring metrics to show areas of higher need—focuses our grant funding and a lot of tree planting."

To help with this work and greater community outreach, the city has established the Youth Forest Council.

"We offer paid internships which provide our youth a pathway to green careers," Halter said. The students in the council work alongside professionals in Austin's urban forestry program, gaining practical knowledge about natural and environmental sciences, and using GIS for urban forestry. Ultimately, the Youth Forest Council helped shape the community tree priority map, providing valuable input about trees, equity, and health. Halter hopes the students start to see how GIS helps people understand complex relationships.

"Of course, we know that climate solutions have the potential to improve the quality of life for all people," Halter says. "But we also know that climate change impacts don't really affect everyone equally.

"And that is at the heart of our plan. We are preserving existing trees and planting new trees, where trees are most needed."

A version of this story by Christopher Thomas titled "How Austin's Map of Trees Helped City Leaders See and Tackle Social Inequities" originally appeared in the *Esri Blog* on May 25, 2021.

MAPPING FOOD AND ESSENTIALS TO GUIDE PEOPLE

Atlanta Community Food Bank; Santa Monica, California; and Cobb County, Georgia

ACROSS THE GLOBE, PEOPLE ARE REELING FROM THE impacts of COVID-19, including the task of procuring food. In the United States, more than 29.8 million students typically receive free and low-cost lunches and breakfasts at school. When these services were not available because of school closures, school officials coordinated to resume meal distributions. When the pandemic began, panic buying depleted grocery stores of basics like canned food and baby formula. As unemployment rates climbed, many households faced desperate situations—the lack of food and funds compounded as parents and children stayed home all day.

In response, city and county governments deployed smart maps and dashboards built with GIS technology to address food needs and shortages—from delivering aid to those who need it most to supporting local businesses with local maps of delivery and takeout options.

Responding to food insecurity

In many cities across the country, food banks took action and used maps to support their mission. The Atlanta Community Food Bank (ACFB) launched its COVID-19 Help Map for families in need, communicating food pickup sites across the Atlanta metro region.

"We estimate 1 in 5 children are in food-insecure households within our 29 service areas," said Heather Moon, public relations coordinator with ACFB. "That means roughly 20 percent of all children do not always know when they'll have their next meal."

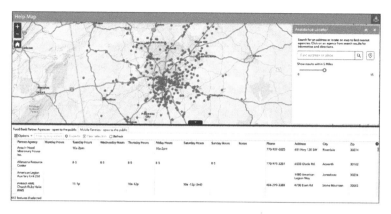

ACFB's COVID-19 Help Map allows users to search for the nearest agencies on a map, providing information on hours of operations and driving directions.

Moon said the food bank works with local public schools to provide meals that students would normally receive in school and groceries families can take home.

The COVID-19 Help Map aggregates data from ACFB's partner agencies. Keeping with the sense of community, the food bank sources its goods from various vendors and encourages other relief organizations and local governments to use the map's feature layers inside their GIS platforms.

The mapped information provides agencies in Atlanta and north Georgia with an authoritative set of points to visit for outreach programs such as the Supplemental Nutrition Assistance Program (SNAP).

Users of the COVID-19 Help Map can navigate subsets of food banks, such as school pantries and meal sites for children 18 and younger. People without access to the map can text a simple message, such as "find food," to a phone number set up by the ACFB to get a reply with the nearest food pantry locations.

ACFB's COVID-19 Help Map had thousands of views, and ACFB had distributed nearly one million pounds of food by the spring of 2021. As the crisis continued through that year, the pace picked up, and the ACFB delivered 250,000 pounds of food per week to 21 public school sites for those affected by school and business closures.

Mapping essential business services

State and county social distancing mandates shuttered many businesses and left people unsure of where to turn for childcare, medical care, or food pantries. Restaurants that remained in business struggled to let customers know they were still open for delivery or takeout orders.

In Santa Monica, California, city officials responded by setting up an online survey for business owners to communicate essential services and restaurants on a map. The city processed survey results twice daily to provide a dynamic up-to-date map. Santa Monica residents can search the map by name or business type.

"Information is shared with the public as quickly as possible since businesses often change the services they offer and their hours of operation," said Zachary Robinson, a GIS analyst for the City of Santa Monica.

The program gained support from local businesses, with more than 20 percent of Santa Monica restaurants completing the survey within the first few days, and more results arrived daily.

"We've heard a lot of positive anecdotal testimony from business owners who were excited to see the city trying to promote their businesses during this pandemic," Robinson said. "The Santa Monica Essential Business Services Map is also being shared widely among residents."

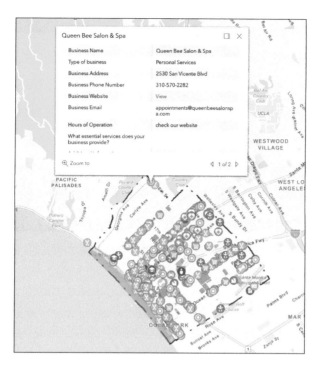

The Santa Monica Essential Business Services Map includes critical categories such as food, medical care, and childcare. Users learn about hours and other details by clicking the icons on the map.

Grocery store inventory for the public by the public

With grocery stores across the country still restocking after the initial panic triggered bulk buying, consumers were left not knowing where to find staples. In Cobb County, Georgia, government staff developed a crowdsourced grocery store inventory survey and map so that shoppers could help others know the availability of essential supplies.

The map shows information about store locations, hours, services, and inventory—organized around key categories such as paper goods, disinfectants, medicines, bread, dairy products, meats, and other popular items. The map lets users know if items are sold out,

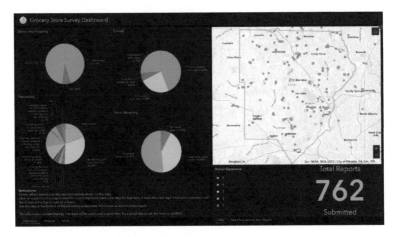

The Grocery Store Survey Dashboard reflects details from crowdsourced reports that citizens record about items in stock at their local grocery store.

and also how crowded the shop is and whether shoppers are practicing social distancing. The data, fueled by updates from residents, changes daily.

The Cobb County Grocery Store Survey Dashboard combines all inputs for an overview of shopper sentiment, what's in stock, and how patterns are trending over time. More than 650 reports were entered within the first week, showing the power of crowdsourcing data in times of crisis.

As more cities and counties deploy maps and dashboards to support residents, they provide much needed support during the COVID-19 crisis for families as well as local businesses.

A version of this story by Christopher Thomas titled "COVID-19: Local Governments Map Food and Essentials to Guide People" originally appeared in the *Esri Blog* on April 17, 2021.

URBAN PARKS PLAY A KEY ROLE IN CURBING INEQUITY

Trust for Public Land

A S THE LOCUS OF MOST ECONOMIC ACTIVITY AND greenhouse gas emissions, cities around the world face the escalating challenge of understanding and managing climate change. Cities are home to half the world's population, and as urbanization increases, so too will climate risk. This risk makes urban strategies to curb emissions all the more critical. Many of these strategies are being built around location-specific insight from data-driven maps and spatial analysis, using GIS.

While cities stand out as primary contributors to climate issues, they are also experiencing some of the biggest impacts of climate change. Most cities lie near coastlines or inland waters, putting them in immediate danger from rising sea levels and extreme weather events.

Taking a more granular view, climate vulnerability across cities is unevenly distributed. Low-income communities suffer more than others, often for deeply rooted historical reasons. As the crisis intensifies, these urban climate inequities will too.

City smarts and smart cities

Leaders in many major US cities are approaching climate risk and social equity issues by rethinking public parks, guided by GIS maps. To that end, the Trust for Public Land (TPL), a San Francisco-based nonprofit that advocates for the creation of parks and preservation of greenspaces, partnered with governments in Boston, Los Angeles, New York, Denver, and other US cities. Ten years ago, TPL launched its Climate-Smart Cities program to address unique climate-related problems in urban areas.

"Parks promote health and improve well-being, build social cohesion when communities come together outdoors and make cities more resilient to climate change," said Lara Miller, TPL's senior GIS project manager.

TPL staff use GIS maps to identify which communities are underserved by parks or identify how many people within a city can reach a park within a 10-minute walk.

Through collaborative work with local governments, TPL helps cities build green infrastructure with an emphasis on climate change resilience. Projects might include new trails and transit lines, shade mitigation to reduce heat islands, parkland and playgrounds that double as flood-reducers, and shoreline parks that protect coastal cities from rising seas.

The climate crisis in New Orleans

Just a few years into recovery from Hurricane Katrina, New Orleans became one of the first TPL Climate-Smart Cities partnerships. Causing more than 1,800 deaths and $125 billion in damage, Katrina was a harbinger of the disruptive weather events that would increase with climate change.

As the city prioritized improving climate resilience, TPL facilitated the creation of "green schoolyards," replacing concrete surfaces prone to flooding with gardens of native plants that absorb rainfall and runoff. Several other interventions, including wetlands restoration and stormwater catchment basins, also addressed flood concerns while increasing open space access for neighborhoods that did not previously have it.

The New Orleans collaboration set a Climate-Smart precedent for encouraging data-driven solutions. TPL developed Climate-Smart

Cities New Orleans, a GIS-based tool to plan and implement projects. Climate-Smart Cities New Orleans stores and integrates location-specific datasets, and projects them as layers on a smart map. For New Orleans, this program included flood data, public health and household income information, and a map layer of green space access across the city.

"We tailor our work based on a city's biggest challenges and goals," said Taj Schottland, the Trust for Public Land's senior climate program manager. "For a city like New Orleans, it's no surprise that the emphasis would be on flooding and absorbing runoff. In other cities, sometimes it's transportation or urban heat islands that emerge as the major priority."

These tools help communities to understand risks, but also guide local action. In New Orleans, "whenever someone proposes a green infrastructure project, the city requires the applicants to use our decision-support tool to justify why they want to do it in a given location," Schottland explained.

The big drawdown

The paradox cities face in the age of climate change is that they are, as the United Nations put it, "the cause of and the solution to" this existential crisis. Even as they compound the problem, cities embody many of the best solutions.

For example, density can promote sustainability. Research suggests that if the United States were to take moderate steps toward promoting housing density and improving transit, by 2030 the country could cut its emissions by a third.

The TPL Climate-Smart Cities program aims to help cities manage the effects of climate change through targeted green infrastructure projects. TPL recently expanded its climate-related activities to help cities understand the part they can play in alleviating the root cause of the climate crisis.

Using the same type of GIS-based tools employed by Climate-Smart Cities, TPL partnered with the Urban Drawdown Initiative—a project of the Urban Sustainability Directors Network—to advance nature-based solutions that capture and store carbon. The program also delivers local health, equity, and economic development benefits for communities.

"We're working with a host of cities, Colorado State University, and other researchers to quantify carbon capture by urban green spaces and then model how different nature-based interventions can increase active carbon capture," said Schottland. "This work is groundbreaking and has the potential to transform our understanding of the role re-greening our cities can play in the climate crisis."

Environmental justice

Intense media coverage of Hurricane Katrina highlighted the equity component of natural disasters—who was impacted more than others—and how the people affected were treated in the disaster's aftermath.

New Orleans' Black residents, faced with disproportionate poverty rates, were less likely to have a means of escaping the city. In the most symbolically stark example, police stopped evacuees from walking over the Mississippi River bridge that connects New Orleans with the much whiter town of Gretna.

In the ensuing years, environmental justice—the equal treatment and involvement of all people in environmental decision making—has become more mainstream.

TPL's Climate-Smart tool has been one way for cities to highlight environmental justice. In a recent collaboration, TPL staff helped Los Angeles planners identify areas of extreme heat within the city to prioritize heat reduction efforts.

Trees soften urban landscapes, provide much needed shade, and offer many more ecosystem services.

"We were able to show the city the census blocks, and the data relating to the census blocks, overlaid with heat islands," Schottland said. "Low-income residents who live in hotter neighborhoods with less tree canopy are less likely to have the resources to pay for air conditioning—or they may be more likely to work outside instead of having an air-conditioned office job. So we want to direct city investments to these neighborhoods and protect those who are experiencing extreme heat."

Maintaining equity

As our society understands the need to tackle social inequity and the climate crisis, the TPL has refined its approach. The TPL builds new GIS-based tools, creates new parks, and helps cities guard against unintended consequences of neighborhood improvement. New green infrastructure such as parks, gardens, and playgrounds can make a neighborhood more desirable, a process researchers call "environmental gentrification."

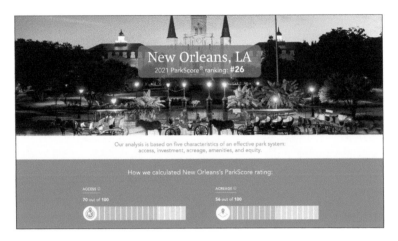

ParkScore for the city of New Orleans.

"There are examples of green space being created in a neighborhood and people being displaced," Miller said. "During planning and project implementation, it is critical we engage community partners to understand gentrification or displacement concerns and bring those concerns into our park development process."

The Climate-Smart Cities tool provides a way to knit together TPL's three pillars of parkland value—health, equity, and climate—into a more expansive view of the modern city.

"The program has always had an equity lens, but our thinking about what equity itself means has evolved and continues to evolve," Miller said. "When it began, we were thinking more in terms of physical access, like who is without air conditioning or those who can't walk to a park within 10-minutes of where they live. Now we can also examine other issues related to equity, like access to information and decision-making."

This evolution continues in another national tool developed by TPL, the ParkScore Index. This index quantifies how well the 100 largest US cities are providing communities with park resources.

Cities are awarded points based on analysis of four important characteristics of an effective park system: acreage, investment, amenities, and access. And a fifth ParkScore characteristic—equity—will better assess how park systems provide equitable access to the health, climate, and community-building benefits of parks and greenspace.

TPL developers also designed ParkServe, another GIS tool that quantifies how well the cities are meeting community needs for greenspace. Using ParkServe, city leaders and park advocates can access TPL's comprehensive database of local parks in nearly 14,000 cities, towns, and communities to guide improvement efforts. This data underpins TPL's engagement with cities.

A version of this story by Jen Van Deusen originally appeared in the *Esri Blog* on February 25, 2021.

USING LOCATION INTELLIGENCE TO MATCH PROGRAMS TO MEMBER NEEDS

YMCA of the USA

THE UNITED STATES IS HOME TO MORE THAN 2,600 branches of the YMCA, serving about 20 million members, or one out of every 16 Americans. It is perhaps the only institution so universal it can be identified by a single letter.

But this ubiquity presents a challenge. Every Y serves communities with different needs. In a diverse society, those needs can be difficult to discern and satisfy.

Adding to this challenge is the Y's federated structure. YMCA of the USA (Y-USA), the national resource office for local YMCAs, provides support for the more than 800 independent nonprofits that operate each facility. "Because we are everywhere, decisions about where to invest can be difficult," said Maria-Alicia Serrano, Y-USA's senior director of Research, Analytics, and Insights.

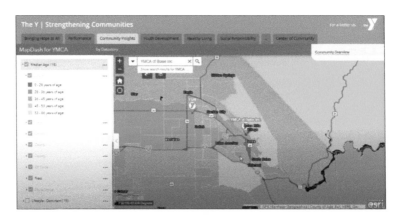

A close look at demographics helped the director of the YMCA in Boise, Idaho, adjust strategy to draw new members.

Boise YMCA gains young members

The director of a YMCA in Boise, Idaho, recently contacted the national office in Chicago for advice on strategic planning. He was looking for ways to increase membership, particularly among seniors.

The Y-USA representative who took the call launched a dashboard and shared his computer screen. The dashboard displayed data from a GIS that Y-USA offers to every location. A map of downtown Boise appeared on the screen, including the Y in question.

A GIS map displayed a 12-minute drive-time around the Boise facility. Highlighting a demographic layer on the map revealed that the dominant age group in the zone was people in their 20s.

The director pointed out that those are areas where developers are converting old industrial buildings into lofts and apartments. "Maybe there are some communities I'm not engaging," he said.

Next, the Y-USA representative turned on a penetration analytics layer of the map. Color-coded areas revealed neighborhoods of lower and higher community engagement with the Y. Two stood out as particularly low. "It looks like I don't need an older-adult strategy," the director said. "I need a Boise Junction and southeast-of-downtown strategy."

Seeing the world anew

The Y adopted its data-driven approach several years ago to better understand how American society was changing, and how these changes should affect the Y's operations on a national, regional, and local level. Part of the implementation challenge has been in helping local staffs feel comfortable using GIS tools.

"At first, some of them are afraid they're going to break it," Serrano said. "And they literally mean that the tool is going to make smoke come out of the back of their computer! So some of what we do is just sitting with them and saying go have fun with it."

The median revenue of a local Y is $3.2 million, and the median number of full-time staff is 12. So, Serrano explained, they need support to access and apply data and insights on their communities. Using GIS, staff can research their communities and follow up on hunches about which programs are needed.

Local insights often resonate on a national level. A Y branch in Minneapolis launched women-only swim classes, partly to respect the cultural and religious mores of the large Somali population in its service area. That prompted the Y to use its GIS to search for other large Somali pockets around the country, displaying them on the map alongside Ys with operating revenues and membership sizes comparable to the Y in Minneapolis. The search yielded 180 other branches nationwide that could benefit by offering similar swim classes.

A far-flung, diverse membership

The original Young Men's Christian Association movement, which began in Great Britain in the mid-19th century, was a reaction to the Industrial Revolution's rampant urbanization. Reformers feared cities were not offering enough wholesome recreational and leisure pursuits.

Although the Y long ago became inclusive, multigenerational, and nondenominational, supporting childhood development is still the core of its mission. Even with more than 2,600 branches, effectively serving the young means partnering with community organizations, especially schools. "To effectively reach these young people, the Y has to go to them and not ask them to come to us," Serrano explained.

For instance, youth programs such as summer day camp and after-school enrichment at the West Cook Y in Oak Park, Illinois, attract strong interest from economically diverse neighborhoods in

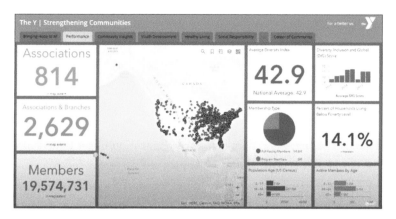

A dashboard view helps Y-USA see the big picture of how it meets its mission.

the western portion of Chicago and its suburbs, known together as Chicagoland. The Y identified 50 schools that could potentially partner with West Cook on these programs.

Rather than approach all 50 schools, the Y's tools mapped out 15-minute walk times for all Y members under 18 in these neighborhoods. The data revealed that partnering with just four schools would allow West Cook to reach 80 percent of these community members.

Crash course in a new community

The Y's data tools can also serve a more generalized function that furthers its mission of community engagement.

Serrano mentioned someone from the Midwest who recently became acting director of a Y branch in the South. He realized that the tool could give him a crash course in the area's social and cultural demographics.

"He wasn't even thinking about using the tool from the perspective of 'how am I going to get more membership?'" Serrano said. "It was more like, 'how am I going to relate to community leaders? Who are they? What's their background?' So it's fascinating to me the various ways it can be used. The biggest aha moment is when people sit down and say, 'Oh, this is what my community looks like.'"

Planning for an uncertain future

The Y's GIS maps and data help the organization better understand the world as it is now while also helping anticipate the future.

"Young people in the communities we serve are seeking services and support different from their parents and grandparents. Generation Z is the first generation for whom the internet was not an invention but a given," said Maria-Alicia Serrano, YMCA of the USA's senior director of Research, Analytics, and Insights, "and the millennial generation represents one in four familial caregivers, a historically large burden."

"So many people will need our services in the coming years that it will be a multigenerational issue," she explained. "What we're faced with is, how do we serve grandparents and the recent high school graduate?"

Y leadership also believes their GIS is nimble and flexible enough to adapt to abrupt changes, such as the COVID-19 pandemic. The organization is in the process of making adjustments to the dashboard to help Ys further stretch operational resources and safeguard the well-being of both employees and community members, including new partnerships.

"These data points are going to open a whole new world in terms of helping Ys understand alternate ways of doing their work," Serrano said. "It's the gift that keeps on giving."

A version of this story by Jen Van Deusen and Emily Swenson originally titled "YMCA Uses Location Intelligence to Match Programs to Member Needs" appeared in the *Esri Blog* on July 28, 2020.

FOOD BANK USES SURVEY SOLUTION TO SUPPORT FOOD DELIVERY

Yolo Food Bank, California

FOR 50 YEARS, YOLO FOOD BANK HAS WORKED TO END hunger and malnutrition in Northern California's Yolo County. The food bank coordinates the storage and distribution of food from a network of manufacturers, grocers, and growers; it includes a vast organizational network of staff, volunteers, and more than 80 non-profit partner organizations. On designated days, residents can pick up food from distribution locations throughout the county.

The coronavirus pandemic increased the demand for food assistance in Yolo County by more than 60 percent; likewise, the need for the food bank's services grew. However, several of the food bank's distribution centers closed because of the crisis, creating a further hardship for some of the county's most vulnerable residents.

Yolo Food Bank needed assistance to deliver food to individuals unable to get to any of the remaining distribution locations. So, the organization enlisted the Yolo County Innovation and Technology Services Department's GIS team. Under a new program, survey-based solutions have helped the food bank sign up families for food delivery and enroll volunteers to deliver food.

Food aid in a pandemic

Yolo Food Bank created the food delivery program to address the needs of people who are unable to leave their homes during the COVID-19 crisis. According to Mike Martinez, IT manager of development and GIS for Yolo County, the food bank approached the county with the problem of understanding how the closure of distribution centers would affect current food distribution programs.

Martinez called in Mary Ellen Rosebrough-Gay, GIS coordinator for Yolo County.

"As things progressed, we realized new distribution centers were not going to be an option for some of our more vulnerable communities. For their health, we decided that we should look at the option of doing deliveries," said Rosebrough-Gay. "Yolo Food Bank asked us for help on how we might achieve that."

The food delivery program is designed specifically for seniors over 65 years old and people with medical needs. People who do not meet these criteria can still receive food, but they must go in person to a distribution center. The Yolo County GIS team created a survey to collect contact information for food recipients and volunteer drivers. The food bank staff also wanted a simple way to calculate how much food they needed to prepare so that they could tell the distribution centers.

A new solution

The GIS team selected ArcGIS Survey123 to create a survey-based solution. The software, which allows users to create, share, and manage surveys, was selected for its ability to do calculations.

For example, the food bank determines how much food the recipient receives, based on their family size. For example, a family of up to four people gets a smaller delivery than a family of five or more. Calculations are also needed for Federal Emergency Management Agency (FEMA) reimbursement, including how many pounds of food are delivered; this enables the food bank to get a dollar amount for all deliveries.

Rosebrough-Gay said she needed a new solution to do calculations, so she turned to Survey123 Connect, a desktop tool. "I started in the Survey123 regular web browser version and moved to Connect because I needed to do calculations."

Staff created two initial surveys: one for enrolling volunteers to deliver food and another one to sign up food recipients. Rosebrough-Gay explained that the recipient sign-up form includes personal information from the applicant such as where they live and what their needs are and asks eligibility questions to determine whether they meet program criteria.

Using ArcGIS® Dashboards, Yolo County created an accompanying dashboard for food recipients that let Rosebrough-Gay filter data according to the day of the week. So, if a user selects Wednesday as a delivery day, the dashboard indicates how many boxes are needed for that day.

The volunteer survey includes screening questions about things like schedule availability and how many pounds the applicant can lift. The food bank effort has 15 University of California, Davis, student and staff volunteers to help manage this, with tasks that include assigning volunteer numbers when a person signs up and checking volunteer availability.

Another helpful feature is the ability to edit existing surveys. The team added more screening questions and asked the volunteers to directly edit their original surveys in Survey123, saving staff time. Previously, when volunteers were asked to complete a new survey, the team would have to put the new and old data together.

The survey form to sign up for delivery was also translated into Spanish and Russian, as many food recipients in the county don't speak English as a first language.

Support for the community

The new Yolo Food Bank delivery program has allowed the organization to provide food to thousands of families. The group also recruit more than 500 volunteers to drive and provide food. The project, which began with two surveys, now includes two apps built

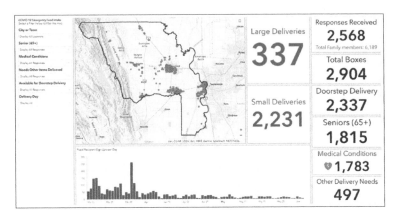

Yolo County COVID-19 emergency food intake dashboard.

with ArcGIS® Web AppBuilder as well as three dashboards and four surveys.

Rosebrough-Gay says that there was a nonstop stream of people being added to the food recipient list. "We've also been very lucky with volunteers and the number of wonderful residents in our county that have stepped up to come and help us with this process," says Rosebrough-Gay. "Even people in other counties have come to help, so it's been great."

Day-to-day operations have been streamlined since the implementation of Survey123. The ability to do calculations with Survey123 has increased efficiency for the GIS team. Recipients enter their family size, and with behind-the-scenes calculations, it can easily be determined how many boxes they will receive and how many pounds of food will be delivered.

"If I didn't have Survey123 doing those calculations for me, I would have to create a spreadsheet, and I'd be manually putting data into it every week. But instead, we can use Survey123 calculations and have it live feed a dashboard we created," said Rosebrough-Gay.

The use of Survey123 has also helped with volunteer recruitment. The screening questions for potential delivery drivers allowed the food bank team to better assess applicants' qualifications and availability, and if a candidate cannot deliver food, the survey populates additional questions to determine how volunteers can help the food bank in other ways. The Survey123 app helps guide volunteer responses.

"This cuts down on phone calls to those volunteers because we already know what days they are available; we already know if they are willing to lift over 45 pounds, because that's how big the boxes are; and things like that," said Rosebrough-Gay. "We still have to talk to the volunteers, but we start with a smaller pool."

The county received positive feedback on the new program setup from Yolo Food Bank, the community, and volunteers. The food bank now has the information it needs to more effectively deliver food to those in need, and volunteers have said that making deliveries is much more efficient with recipients' exact address and location details available.

Rosebrough-Gay said she's happy to be part of a program that has helped county residents.

"A lot of times, as a county employee, you don't know how your work is affecting people. With the food delivery program, we can see how many people we are helping. The GIS team can see that we have been able to deliver over 34,000 deliveries to some of our most vulnerable populations," says Rosebrough-Gay. "And we're able to keep those people at home and deliver the food they need. It feels really good to be able to be a small part of that."

A version of this story by Cassandra Perez originally titled "California Food Bank Uses Survey Solution to Support Food Delivery during the Pandemic" appeared on Esri.com in 2020.

PART 4

ACCESS TO CARE

SOCIAL, RACIAL, ECONOMIC, AND PHYSICAL FACTORS can all act as barriers to restrict access to health care and services and prevent improved health outcomes. Considering these impediments helps providers, health centers, and preventive care centers match their services to community needs. Location intelligence helps providers identify gaps in access to services. Fusing location intelligence with demographics, GIS helps providers respond more specifically to the needs of diverse communities.

Meet network adequacy requirements

Network adequacy standards ensure that health plan members can have reasonable access to health services. Facility siting, mobile and e-health options, and the availability of in-network providers all improve geographic access to health services. Location intelligence can improve health networks to help providers identify people with special needs and allow organizations to lessen the impacts of the social determinants of health. The US Department of Human Services defines social determinants of health as the environmental conditions "where people are born, live, work, play, worship, and age that affect a wide range of health functioning and quality-of-life outcome and risks."

Connect people to services

Data is the currency for analyzing how well health-related services are being placed in the paths of people and access is being improved. GIS provides a framework for matching health services to population needs, addressing disparities in accessibility, delivering health organizations with key metrics, and informing and educating patients.

Ensure universal health coverage

To achieve universal health coverage by 2030 as part of the United Nations Sustainable Development Goals, individuals and communities around the world must access the full spectrum of essential health services—from illness prevention and health promotion to treatment and recovery care. GIS can help identify where to make the most of new and existing resources.

GIS in action

This section will look at real-life stories about how health and human services organizations use GIS to improve routing and locate childcare.

CITYWIDE HEALTH INITIATIVE IMPROVES LIVES

Rancho Cucamonga, California

I N 2008, THE CITY OF RANCHO CUCAMONGA, CALIFORNIA, faced the sad fact that the health of its citizens was declining. Obesity and diabetes were on the rise, a phenomenon that is occurring in many cities and towns across the United States. In response, the city decided to develop the Healthy RC initiative. It began as a special project in which staff reviewed local health rates and compared that information with county data. The findings indicated that half the city's adult population was overweight or obese.

City staff knew that success for the initiative would require collaboration across all departments, community-based organizations, businesses, cultures, and public institutions. They started by getting the departments together and asking what each was doing to improve community health and how their processes and workflows could change to support the Healthy RC initiative.

Through interdepartmental collaboration and comprehensive community engagement, staff developed a strategy for Healthy RC, asking residents, businesses, schools, faith communities, and other stakeholders how to improve the health of individuals, families, and neighborhoods. The strategic plan included eight health priorities:

- Healthy eating and active living
- Community connections and safety
- Education and family support
- Mental health
- Economic development

- Clean environment

- Healthy aging

- Disaster resiliency

Staff recognized that location intelligence was essential to understanding the health of their community and finding solutions that improved the lives of citizens.

The challenge

The primary goal of Healthy RC aimed to improve the lives of Rancho Cucamonga's citizens. While the city knew that health problems were rising, it needed more specific information—more data—to support good decisions and take appropriate action.

In most cities, the location and composition of neighborhoods can represent significant disparities between incomes and access to nutritious food and health care. Citizens living in one neighborhood may have limited or no access to amenities such as parks, farmers markets, and grocery stores, while nearby, other people have opportunities for healthy lifestyles. Whereas one area could be seen as dangerous or unsafe for children to walk to school, another may have community gardens that are not available to people living close by. Even zoning, which determines the type of permitted land use, can contribute to negative health effects for some of the population.

For the City of Rancho Cucamonga, perhaps the biggest hurdle to getting needed data was in reaching out and engaging the citizens themselves. To achieve the granularity of localized data, staff needed to do the in-person health surveys necessary for understanding the lives of citizens in different parts of the city. The data would serve as a baseline for verifying assumptions about neighborhood characteristics and, ultimately, help develop targeted strategies to improve health outcomes.

Policy in Action

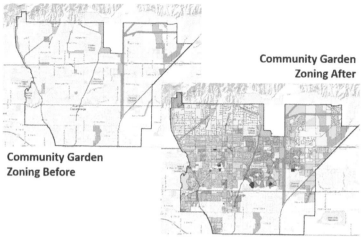

Community Garden
Zoning After

Community Garden
Zoning Before

Community Garden Zoning Policy before-and-after maps. Yellow and orange shaded areas on the after map represent community gardens added as part of the initiative.

To meet the goals and objectives of Healthy RC, the city had to consider all these factors. The Healthy RC team needed a way to bring the data together in one place so it could be analyzed and the findings presented to decision-makers and the public.

The solution

Using ArcGIS, the Healthy RC team members first analyzed neighborhood-level data and identified the most disadvantaged areas, where health issues were greatest and healthy options were scarcest. Next, they compared those areas with Rancho Cucamonga's overall poverty and median income data to better understand the disparities between different parts of the city.

To establish a baseline, team members created a paper-based survey to use as they went door to door to interview people. They also held community events, went to city parks and trails, and deployed an online survey. This way, the team could connect with people wherever they were.

With this initial set of new information entered in ArcGIS, staff quickly identified food deserts—areas where it is difficult to purchase affordable, healthy food. Using maps, team members could see where sidewalks, grocery stores, and parks already existed or were needed. To learn which areas were considered unsafe, they connected the data with the survey information, crime statistics, and community stories.

Through collaboration, citizens used maps to identify places that were unsafe for their children to walk or play in. Residents also communicated where they would most like to see a community garden or a farmer's market.

Bringing this information together in ArcGIS allowed the city to investigate how current zoning affected neighborhood health and how changes to the zoning could benefit different neighborhoods.

The more information the team gathered, the faster the city could make informed decisions and take meaningful action. Additionally, the Healthy RC team began using smart technology to speed data collection. Fieldworkers used ArcGIS Survey123 on smart tablets to collect data more accurately, which allowed them to identify issues at the census tract level. By going digital, the team expedited companion projects, such as the Safe Route to School initiative, which was designed to work with children to show where safer crosswalks and lower speed limits were needed.

The results

Ten years later, the Healthy RC team surveyed the community again, hoping to see improvements from the implemented programs. The positive results of Healthy RC are staggering.

In the first couple years, many issues were identified and policies changed. With community input, the city changed zoning so that the farmers markets and community gardens became more accessible to all neighborhoods. The $1,500 conditional use permit fee for those amenities was waived. These were just two policy changes among many, including completing streets, setting nutrition and beverage standards, and designating smoke-free areas. The city added several infrastructure enhancements, including new sidewalks, bike lanes, and parks.

As stated in a recent report, the survey showed that childhood obesity decreased by 13 percent; the number of overweight students decreased by 7 percent; and the incidence of heart disease, diabetes, and cancer decreased by 20, 14, and 21 percent, respectively.

The city used GIS technology to identify areas where government involvement was needed and make the intervention relevant and engaging to a community.

Staff used ArcGIS® StoryMaps℠ to help people visualize and understand the policy changes. It included maps depicting areas before and after zoning changes. It also contained images showing how the addition of sidewalks and parks was changing the built environment.

Currently, one-quarter of Rancho Cucamonga's population is actively engaged in the Healthy RC program. The city has plans for expanding it citywide through community engagement to also address health equity. The revised program will include strategies to ensure adequate representation of the city's diverse population for shaping future policy, projects, and services.

A version of this story by Natalie Carter titled "Citywide Health Initiative Improves Lives in Rancho Cucamonga" originally appeared on Esri.com in 2019.

HELPING ESSENTIAL WORKERS LOCATE CHILDCARE FACILITIES

State of Washington

WHILE MANY PARENTS BEGAN TO JUGGLE REMOTE WORK with their children's remote learning during the COVID-19 pandemic, many essential workers struggled to find childcare as they reported to work each day.

Health care professionals, first responders, grocery store workers, public service employees, gas station attendants, bankers, and others considered essential risked their health to be at work. For essential workers with children, the situation was more challenging, because many schools and most childcare facilities closed.

In Washington, one of the US states hardest hit by COVID-19, the governor's office mandated that schools provide childcare for essential workers. However, not every school participated, so it was difficult for parents to find which facilities were open.

Staff at the state's Washington Technology Solutions' Office of the Chief Information Officer (OCIO) saw this disconnect and wanted to help.

"Essential workers are risking their lives every day," said Joanne Markert, GIS coordinator for the OCIO. "The least we can do is help them to easily find childcare."

Markert and the geospatial team created an online, map-based dashboard to show which facilities were open for childcare. Essential workers could use the interactive dashboard to select a school district and quickly see available facilities. They could also click the individual facility to learn more details, such as the name, address, email and/or phone contact information, county location, and hours.

The Washington State public dashboard used GIS to help essential workers find the childcare facilities they needed.

Building the interactive map required collaboration across agencies and the use of GIS technology from Esri. The team at OCIO used ArcGIS Dashboards to create the interactive map. Staff at the Office of the Superintendent of Public Instruction provided the underlying GIS data on schools and school districts. The State Emergency Operations Center provided additional information about hours of operation, specific email addresses or phone numbers to contact, and alternative sites such as Boys & Girls Clubs working in partnership with the schools.

The public accessed the dashboard on the Washington Geospatial Open Data Portal. As with everything during these unpredictable times, the facilities and their availability constantly changed, so the dashboard provided workers one authoritative place for the most current status updates.

"We hope that this dashboard alleviates a bit of stress for these folks during such a traumatic time," said Markert.

A version of this story by Megan Martinez titled "Washington State Helps Essential Workers Locate Childcare Facilities" originally appeared on Esri.com in 2020.

BETTER ROUTING LEADS TO BETTER HEALTH CARE

Health services company

F OR YEARS, ONE HEALTH SERVICES COMPANY'S TEAM OF nurse practitioners and physician assistants had passed each other on highways and city streets. They made home visits to members of a health program focused on providing preventive care. Nationwide, more than 1,000 practitioners carried out in-home medical exams for the plan's members, racking up millions of car miles a year en route.

The zig-zagging drained the team's efficiency. Clinicians found themselves bumping into one another in apartment building stairwells, having come from opposite ends of town to cover two different members in the same complex. Then they headed back out in opposite directions.

"The [clinicians] had often noted the challenges of passing each other during the day," explained the program's regional director of clinical operations, who along with colleagues shared the company's story anonymously so they could talk more freely about it.

The company faced a modern-day version of the classic traveling salesman problem: how does a business deliver services or goods across a geographic area as efficiently as possible? Multiply that question by the hundreds of thousands of member visits the program completed in 2020 alone, and it becomes a burning issue to solve and a challenge ripe for location intelligence.

Route optimization could improve member, employee satisfaction

The routing challenges clinicians faced each day weren't just inconvenient; they took valuable time. Nurse practitioners and physician assistants often felt rushed as they moved between appointments and were limited in the number of cases they could cover in a typical 10-hour workday.

That's a familiar challenge for executives in sectors as diverse as retail, restaurants, home services, and logistics. Before the COVID-19 pandemic, consumers already embraced home delivery in large numbers. Now, store closures and a stay-at-home mentality accelerated the shift, sending retail e-commerce sales up 28 percent in 2020.

Businesses that rely on location intelligence are positioned to thrive in this new reality. An IDC Canada survey showed that organizations with mature location analysis practices saw a 25 percent improvement across key business metrics over a two-year period. Whether a company is moving people, transporting goods, or providing services, the GIS technology that creates location intelligence is often at the heart of efficiency improvements.

At the health services company, the leadership team knew that smarter routing of home visits would do more than decrease costs. If done right, it could give clinicians more time with members, reduce cancellations, establish more predictable appointment times, and improve already strong member satisfaction scores.

The move could also have a positive effect on an employee's work-life balance. Well-trained clinicians were already in high-demand before COVID-19. Overall employment of nurse practitioners, anesthetists, and midwives has been projected to grow by 45 percent from 2019 to 2029. By creating better member and practitioner experiences, providers can cultivate a working environment that helps attract and retain top talent in tight markets.

The question for company executives wasn't whether to create a more efficient network to support exceptional patient care and employee experiences, but how.

Using data to transform human experiences for the better

For the health services company, the solution had to match the right technology with an understanding of exactly what the program is designed to deliver: proactive, preventive health care.

"We are not a delivery product," said the company's vice president of analytics. "We focus on human interactions, so the team studied how to improve logistics without compromising an exceptional member experience."

The result was a pilot route optimization program that the company rolled out in Texas in late summer 2019. Focused on the program's logistics strategy, the pilot program aimed to improve the bottom and front lines for several dozen local clinicians and the members they serve.

Working together, the analytics and program teams developed a process to feed scheduling data into GIS route optimization algorithms, which generated the most efficient distribution of appointments and routes, including which clinicians should visit each member.

For the program's frontline workers, the impact was huge. Dispatchers sent optimized route plans to a clinician's tablet or smartphone, and the clinician benefited from the efficiency of smarter routes and shorter drive times.

Still, while GIS excelled at planning everything from John Deere's dealer locations to maintenance work on FedEx planes, the home visits program had traits that traditional use cases did not. The team needed enough flexibility in its route optimization solution to satisfy all the nuances of member care.

Some members prefer either a male or female clinician. Others request the same clinician they saw during the last visit, or one who speaks a certain language. Those prerequisites culled about 25 percent of possible appointments from the pilot program. The analytics team used GIS algorithms to analyze the rest.

That required tweaks like holding off on announcing which nurse practitioner or physician assistant would cover an appointment until much closer to the date.

"This allowed the GIS algorithms to wait for all—or at least most—appointments to be scheduled before committing the assigned APCs," explained the analytics team's data scientist and GIS analyst. "That way the algorithms had as much information as possible on schedule adjustments and cancellations before making a decision."

During the pilot program, analysts manually verified the proposed routes to ensure that no patient preference, however subtle, was overlooked—a process that would eventually need to be automated in order to scale.

Starting local, thinking national, and building momentum

By February 2020, the route optimization pilot program had helped program managers reassign more than 300 appointments along 200 routes, and the GIS-optimized schedules had saved clinicians more than 2,000 miles and 45 driving hours in just five months. That amounted to a 10 percent decrease in drive time across the routes changed and a 12 percent drop in drive distance—a substantial cost savings for the relatively small initial group.

The regional program director says that when routes are more efficient and days run smoother, clinicians provide an even higher level of service for a program that is already well regarded by members. When members can consistently rely on predicted arrival times and see their clinicians working without rushing, it makes a real difference in the member's day. It also empowers clinicians to focus not simply on making the appointment but making the most of it.

The nurse practitioners and physician assistants say shorter routes between visits allow them to repurpose saved minutes into member care, according to the director. "They can spend that time with our members ensuring a high-quality exam, teaching, and addressing any of the member's needs."

That supports a better working environment at a time when health care professionals are stretched thin. In the director's words, the optimization program is a way "to help our employees have a better work-life balance and then give that time back to our members."

The results became a promising precursor to the company's next business challenge: how far could they extend those savings and improvements by applying GIS across thousands of clinicians, and millions of appointments, nationwide?

As the pilot program proved itself, the world shifted gears

Buoyed by the pilot's progress, the company was poised to expand the program nationally. Then the world ground to a halt. The company primarily serves Medicare members, and their age made them particularly vulnerable to COVID-19. In deference to safety protocols, program directors temporarily prioritized virtual appointments over in-home visits.

The analytics team used the time to study the results from the pilot stage and ensure they were ready to expand GIS route optimization across the program nationally, which they did in Q4 of 2020. The analytics vice president said they're now well positioned to tackle what many call a care gap that's widening under the weight of a strained health care system.

The team created processes that will make GIS-optimized routes feasible on a much larger scale. And they're incorporating stakeholder feedback at every step. Most importantly, they're working to keep their members and clinicians at the heart of the location intelligence solution.

"This program has been proven to reduce hospitalizations and increase visits to the primary care provider, closing the loop to ensure once we leave, that journey continues onto that healthier life," the regional director said. "Being out there in the home is really important and relevant right now."

And it's likely to become even more relevant as the company expands its use of location intelligence to increase levels of member care.

A version of this story by Este Geraghty originally appeared in *WhereNext* on May 11, 2021.

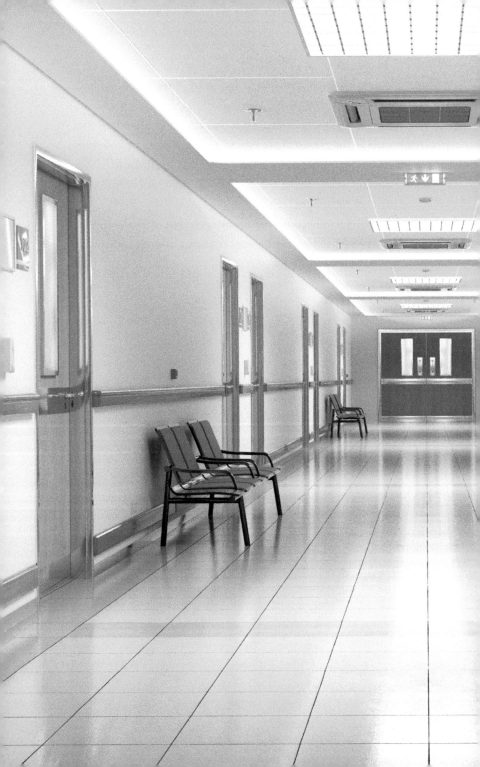

PART 5

STRATEGIC PLANNING

L OCATION DATA IS ESSENTIAL IN HELPING HEALTH CARE
and health systems professionals create more effective strategic plans. This enhanced health care strategic planning involves allocating resources; responding in real time; adjusting long-term goals while considering environmental, physical, and demographic trends; and improving crisis management. Through GIS technology, information can be collected in real time and fed back into operations dashboards. Adding the location component to data and enterprise systems enables a patientcentric approach to improving core functions, managing assets, enhancing communication, and tackling issues holistically.

Understand demographic trends

Knowing a community's demographic and lifestyle makeup sets the foundation for creating a sustainable operational plan. Location intelligence paves the way for recognizing where the existing patient base resides, identifying untapped neighborhood segments, planning a service mix, and preparing for emergencies. GIS shows a baseline of the community as it is today and supports the forecasting required for sustained growth.

Site selection and facilities management

Bringing health care directly to neighborhoods in need is a departure from a "build it and they will come" mind-set. GIS helps plan for future health care facilities based on patient needs, mobility, and at-risk populations, and assists with target outreach efforts. From mobile nursing services to outpatient clinics, GIS can place health services where they need to be. GIS can also be used indoors to optimize space planning, facility management, and navigation within a health facility.

Analyze service gaps

The key to identifying service gaps is a thorough understanding of patient mix, competition, demographic shifts, and outlying populations. GIS provides the ability to analyze service and market area, determine where it succeeds, and pinpoint underserved neighborhoods. Dashboards can map performance goals and present opportunities to improve service levels.

Enhance market development

GIS gives management teams unprecedented access to data and business intelligence to assess current markets, improve community awareness of offerings, and enter new health care industry segments. Location intelligence offers a competitive advantage in analyzing where patients are coming from and, more importantly, where they are not. Targeting markets based on location and market segmentation data can lead to unprecedented growth.

Meet regulatory requirements

Today's organizations operate in an environment where they must meet reporting requirements. GIS can help align with standards and validate data, all while keeping patient privacy at the forefront. GIS assists in performing needs assessments and inspections, and it helps meet reporting requirements through enhanced demographic information, data visualization and analysis, and advanced monitoring.

GIS in action

This section will look at real-life stories about how health and human services organizations use GIS to improve patient outcomes, help schools navigate the pandemic, and more.

CONNECTING LOCATION TO BETTER PATIENT OUTCOMES

Loma Linda University Health in California

LOMA LINDA UNIVERSITY HEALTH (LLUH) IS RECOGNIZED AS one of the top academic medical centers in Southern California. Its operations include six hospitals, a physician practice corporation, remote clinics in the western United States, and affiliates around the globe. The center is known and respected for advanced technology, service-oriented medical care, and education. It is recognized worldwide for its revolutionary efforts including the first proton unit used for cancer treatments and the first infant heart transplant. The Loma Linda University School of Public Health also pioneered health GIS research, successfully using location for improved health outcomes. LLUH's focus on location intelligence is now spreading to the operations of its facilities.

The challenge

Treating more than 1.5 million outpatients every year prompts LLUH to continue to look for ways of improving patient care through a digitally driven insight. The hospitals and ambulatory clinics rely on Epic's electronic health records (EHR) system to manage their patient data. In a parallel system of record, they rely on ArcGIS to maintain, manage, and analyze their location data, which enables patients to easily navigate to the correct buildings, parking lots, and offices for their appointments. After some careful thought as to how best to bring about a more holistic view of their patients, management staff recognized the potential of gaining new insights by integrating these two systems. They realized that using this location data would transform their operations and, in turn, improve patient experience.

The solution

Critical to Epic's functionality is its Admission, Discharge, Transfer (ADT) module, which houses the location and status of every patient. The medical staff were tracking all this critical information, but they had no way to visualize and analyze it. Combining the Epic EHR data with the patient location data in the GIS quickly transformed the information into actionable information, such as rapidly identifying hot spots and emerging trends and improving coordination of work across departments.

By integrating their two existing enterprise systems, Epic and ArcGIS, LLUH staff can now quickly and easily see the locations of their patients. By connecting the ADT feed with ArcGIS, staff can now quickly improve decision-making through a real-time operations dashboard of the status of their patients. The new workflow doesn't require any additional time or effort for staff. They simply enter the patient data as they have in the past; the only difference

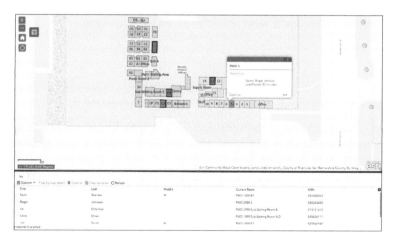

Combining Epic EHR data with ArcGIS allows LLUH providers to spatially assess critical patient data. This interactive map shows rooms in a medical facility; clicking on a room displays a pop-up with patient information.

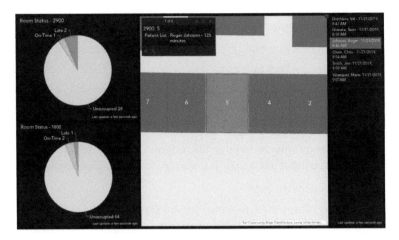

By connecting the Epic data feed with ArcGIS, LLUH providers now have a complete real-time view of their patient flow, enabling them to identify increased wait times and respond accordingly.

is that the data is now being displayed visually. They can now easily see what's happening with their patient load in real time, identify long wait times or trouble spots, and respond accordingly.

The results

The immediate value is getting a comprehensive view of the patient flow through the clinic. Over time, they will analyze the data to identify trends and patterns, which can then be used to make better decisions and improve the patient experience. As Bert Chancellor, the executive director of information services at Loma Linda University Health, explained, "The real value is not just visualizing something; everyone can do that. The real value is to have the knowledge and ability to be impactful over a period of time."

With strong support from leadership, LLUH is planning to roll out this integration to additional clinics in the future. As Chancellor explained, "Integrating enterprise platforms is transformative.

Enhancing electronic health records with location analytics improves the patient experience, resulting in data-driven decision-making, and provides extra insight into data."

A version of this story by Megan Martinez titled "Loma Linda University Health Continues to Innovate by Connecting Location to Better Patient Outcomes" originally appeared on Esri.com in 2019.

IMPROVING COMMUNITY HEALTH

Providence St. Joseph Health

WITH A HERITAGE OF IDENTIFYING AND SERVING community need, Providence St. Joseph Health conducted Community Health Needs Assessments (CHNAs) long before they were required by the Affordable Care Act. In partnership with Esri, Providence St. Joseph Health made the CHNA a core component of organizational strategy. Since taking a location-centric approach to its community benefit efforts, Providence St. Joseph Health—a national, Catholic, not-for-profit health system—gained unparalleled insights that allow it to deliver better service to its patients and communities at large.

Rather than producing static documents every three years, Providence St. Joseph Health wanted to find a more effective way to fulfill the CHNA requirements and make itself a community-based resource. Understanding the needs of each community is a critical step in achieving Providence's vision of "health for a better world." In 2018, Providence began modernizing its CHNA process to ensure inclusive, mixed-methods approaches that promote collaboration throughout the system's 51 hospitals across seven states. These results inform outcomes, resource allocation, and hospital strategy to address prioritized community health needs.

Providence St. Joseph Health has teamed with Esri to reimagine the CHNA and overcome the barriers of the past. Together, they're developing a standardized approach to the CHNA that focuses on people and their environments, how they are related, and what actions will move the needle toward improved health outcomes. The companies' new, digital CHNA will provide a greater understanding and serve as an operational tool for prioritizing community-based efforts and resource allocation.

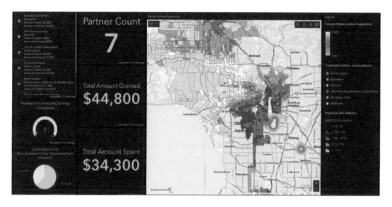

With ArcGIS, Providence St. Joseph Health can efficiently visualize the vast quantities of data being gathered through their engagement with community partners and residents. This dashboard shows the deployment of resources to combat diabetes in Los Angeles.

Using ArcGIS® Business Analyst™ software and data, Providence St. Joseph Health will analyze demographic and location data to create actionable intelligence. This capability will help ensure that health equity is at the forefront of Providence's efforts and drive community investment strategies from a place-based perspective. The map-based digital CHNA will make vast quantities of data immediately understandable and far more accessible to the public. Implementing Esri's community engagement software, ArcGIS® Hub℠, will enable Providence to turn the CHNA into an interactive resource to foster bidirectional communication and facilitate sharing and collaboration with patients and community members.

This innovative CHNA will enable users to view local issues—such as homelessness, chronic disease, and substance abuse—in a real-time way and easily provide feedback and share this information with others in the community. This capability will allow Providence to be a leader in implementing place-based approaches to addressing social determinants of health and monitoring their impact with community-level data.

Another significant benefit for Providence St. Joseph Health is the partnerships it has already begun to form and strengthen. Many of the county organizations Providence partners with through the CHNA process also use ArcGIS. The use of this software allows them to cross-pollinate their vital databases and relationships through the common language of GIS.

Dora Barilla, vice president for community health investment at Providence St. Joseph Health, said "Through this effort, it is our belief that the CHNA will come alive, moving from a binder on the shelf to a dynamic, interactive, and strategic tool that contributes to and catalyzes our communities to achieve health for a better world."

A version of this story by Megan Martinez titled "Seeing What Others Can't: Improving Community Health and Health Equity" originally appeared on Esri.com in 2019.

HELPING SCHOOLS NAVIGATE THE PANDEMIC

Riverview School District, Pennsylvania

THE COVID-19 PANDEMIC DISRUPTED EVERY ASPECT OF American life. During this pandemic, school superintendents faced unique challenges. Not only must they ensure the best educational outcomes for students, they must also protect the health of students, teachers, and communities.

"In addition to making the usual safety, operational, and educational decisions, superintendents are now expected to assess the current landscape in order to make critical choices regarding public health," said Neil English, the superintendent of Riverview School District in Pennsylvania. "Navigating the health crisis is unchartered territory, and school leaders are tirelessly managing the confluence of local, state, and federal mandates; CDC [Centers for Disease Control and Prevention] guidelines; and local health metrics to assess risk and make informed decisions regarding the health and safety of the students and their families."

Like other school districts in the United States, Riverview was flying blind. However, Esri partner Epistemix helped the district make better-informed policy decisions. Epistemix used its epidemiological simulation platform to model the local, specific health impacts of the response and opening strategies under consideration.

Using a statistically accurate digital twin of the US population that individually represents every student, teacher, and resident within the district and surrounding areas, Epistemix calibrated the COVID-19 model to the evolving epidemic with realistic social dynamics, using ArcPy geoprocessing tools. The results were imported into a hosted layer in ArcGIS® Online with additional

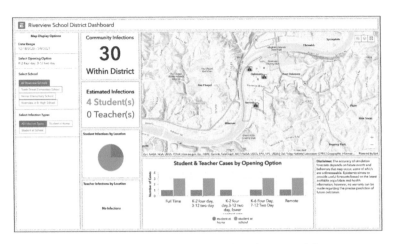

Epistemix calibrated the COVID-19 model with realistic social dynamics, imported the results into a hosted layer in ArcGIS Online, and presented them to the district.

data from ArcGIS® Living Atlas of the World. The results were presented to the district using ArcGIS. A dashboard gives English and the school board the information needed to lead the district through this pandemic, including specific estimates for student, teacher, and community infection rates for four different opening strategies with breakdowns for individual schools within the district.

Riverview isn't flying blind anymore. "Epistemix helps to lead the way by providing much-needed projection and modeling data that can help superintendents make more informed decisions," said English. "And more informed decisions are a valuable commodity in the wake of COVID-19."

A version of this story originally appeared in the Spring 2021 issue of ArcNews.

NEXT STEPS

The geographic approach to health and human services

HEALTH AND HUMAN SERVICES PROFESSIONALS worldwide rely on GIS to understand coverage gaps in the populations they hope to reach, identify opportunities to improve workflows, plan for and mitigate against unforeseen events, and more. Applying GIS maps and analytics leads to improved health outcomes, increased accessibility to health care, and healthier communities. Following are some recommended steps to help you get started with GIS.

Identify foundational data

Gather and map your foundational data in your area. These layers include the basic infrastructure and administrative areas:

- Administrative boundaries (city and country boundaries, police districts and precincts, fire districts, etc.)
- Population and demographics
- Public safety infrastructure (police stations, fire stations, etc.)
- Structures and structure type (single-family homes, multifamily units, commercial use, etc.)
- Major facilities and landmarks (schools, malls, places of worship, parks, stadiums, etc.)
- Health infrastructure (hospitals, clinics, assisted living facilities, etc.)

- Shelters

- Roads

- Bridges

- Dams

- Utilities

- Communications infrastructure

- Water features (lakes, streams, rivers, etc.)

- Parcels

- Addresses

Include hazard-specific data that is relevant for your area of interest. If you are unsure of what hazards present the greatest risk in your area, tools such as the Federal Emergency Management Agency's Resilience Analysis and Planning Tool (RAPT) in the United States can help you assess priorities.

Add ready-to-use, curated content from ArcGIS Living Atlas of the World. ArcGIS Living Atlas contains several ready-to-use live feeds that provide dynamic, real-time information that can be used in addition to your local data:

- Weather feeds

- Disaster feeds

- Earth observation feeds

- Multispectral feeds

Discover more live feeds in ArcGIS Living Atlas at links.esri.com /atlas_live.

Also consider adding in real-time services for additional situational awareness:

- National Shelter System

- World Traffic Service

- WAZE Traffic

Identify data gaps

After collecting and organizing the base data, you can assess it and identify data gaps. The data drill is a multi-organization exercise used to gain insight into how a community collectively thinks about, manages, shares, and uses data during a crisis.

Data drills are developed and conducted based on operational challenges involving data and are a valuable tool for disaster preparedness. Data drills can be designed around a specific scenario relevant to your community such as an infectious disease outbreak, fire, or earthquake to ensure you are planning for all data needs.

Here are a few things to consider in your data drill. Once completed, you can develop a plan to collect or create data where required based on these suggestions:

- Detail your organization-specific operational workflows and use cases based on the scenario.

- Identify the relevant decisions and determine what datasets, including metadata and data dictionaries, support those key decisions.

- Look next at your interagency workflows based on the scenario and identify key decisions and need for data support.

- For each data point, identify the responsible organization contacts, roles, and responsibilities for this dataset.

- Identify if you will need any data sharing agreements between partners and start collecting and sharing the data identified in this drill.

Create and share maps

Once you've located the data sources you need, you can create a variety of maps to help keep your community informed and healthy, starting with these mapping tasks:

- **Map capacity:** Use the capabilities of GIS to map your facilities, infrastructure, employees or citizens, medical resources, public safety resources, equipment, goods, and services to understand and improve access.

- **Map hazards:** Prepare for public health crises by creating maps showing the location of hazards that could potentially impact your community, such as areas with high crime rates, seismic vulnerability, flood zones, and wildfire risk.

- **Map vulnerable populations:** More specifically mapping social vulnerability, age, and other factors helps you monitor at-risk groups and regions you serve, and which could be even more impacted in a public health crisis.

- **Share maps and plans:** Sharing your maps and plans with your community provides transparency, increases equity, helps citizens understand risks, and improves community preparedness.

- **Follow best practices:** Ensure that your maps and apps are ready to handle the load from the public and media during a crisis and that your GIS environment is ready for the next response.

Learn by doing

Hands-on learning will strengthen your understanding of GIS and how it can be used to improve health and human services. Learn ArcGIS is a collection of free online story-driven lessons that allow you to experience GIS when it is applied to real-life problems. Learn ArcGIS includes these and other lessons applicable to health and human services:

- **Introduction to ArcGIS Online.** Get started with web mapping with ArcGIS Online.

- **Integrate maps, apps, and scenes to tell a story.** Share information about earthquake risk using maps, apps, and scenes.

- **Site a new hospital.** Find possible locations for a new hospital based on spatial criteria.

- **Analyze medical facilities for expansion.** Build a web app to access demographic data for your clinic sites using the ArcGIS Business Analyst widget.

- **Track virus spread with ArcGIS® Insights℠.** Use link analysis to trace contacts as a virus spreads through a community.

- **Investigate prescribed drugs.** Fight drug abuse by identifying suspicious prescription trends.

- **Help end homelessness.** Build data-driven solutions for combating homelessness.

- **Identify landslide risk areas.** Analyze a map to predict future mud flows in rain-soaked Colorado.

You can try these and other Learn ArcGIS lessons for health and human services at learn.arcgis.com.

Get there faster with GIS templates

ArcGIS Solutions for Health and Human Services reduce the time it takes to deploy location-based solutions in your organization. You can use these solutions, or templates, to take action that helps prevent the spread of diseases, analyzes health care patterns and trends, maintains a clean environment, encourages healthy behaviors, and ensures access to health care services.

ArcGIS® Solutions for Health and Human Services include these templates:

- **Community Health Assessment.** The Community Health Assessment solution can be used to collect community health information, at the household-level, required for disaster response or health action plans.

- **Community Health Outreach.** Community Health Outreach can be used to communicate health initiatives, increase participation in key initiatives, and ensure the public knows where they can locate health services.

- **Homeless Point-in-Time Counts.** Homeless Point-in-Time Counts can be used to conduct point-in-time counts of sheltered and unsheltered homeless individuals.

- **Homeless Risk Reduction.** Homeless Risk Reduction can be used to identify where regular reports of homeless activity may originate and address regular reports of homelessness activity in a community.

- **Homeless Outreach.** Homeless Outreach can be used to educate the public and help individuals experiencing homelessness find available resources.

- **Opioid Epidemic.** Opioid Epidemic can be used to communicate the severity of the opioid epidemic, promote treatment alternatives, and understand the effectiveness of response activities.

- **Mosquito Service Requests.** Mosquito Service Requests can be used to solicit reports of mosquito activity and locations of potential breeding sites from the general public.

- **Mosquito Treatments.** Mosquito Treatments can be used to manage adulticiding and larviciding activities and share treatment plans with the general public.

- **Mosquito Surveillance.** Mosquito Surveillance can be used to monitor mosquito populations and track the location of positive vector-borne disease test results.

- **Restaurant Inspections.** Restaurant Inspections can be used to inspect restaurants or food service establishments and share inspection information with the general public.

- **Performance Management.** Performance Management can be used to monitor key performance metrics and communicate progress made on strategic outcomes to the general public and other interested stakeholders.

Learn more

For additional resources and links to live examples, visit the book web page:

go.esri.com/mch-resources

CONTRIBUTORS

Matt Ball
Jim Baumann
Chris Chiappinelli
Keith Mann
Amen Ra Mashariki
Monica Pratt
Citabria Stevens
Carla Wheeler

ABOUT ESRI PRESS

A T ESRI PRESS, OUR MISSION IS TO INFORM, INSPIRE, AND teach professionals, students, educators, and the public about GIS by developing print and digital publications. Our goal is to increase the adoption of ArcGIS and to support the vision and brand of Esri. We strive to be the leader in publishing great GIS books, and we are dedicated to improving the work and lives of our global community of users, authors, and colleagues.

Acquisitions

Stacy Krieg
Claudia Naber
Alycia Tornetta
Craig Carpenter
Jenefer Shute

Editorial

Carolyn Schatz
Mark Henry
David Oberman

Production

Monica McGregor
Victoria Roberts

Marketing

Mike Livingston
Sasha Gallardo
Beth Bauler

Contributors

Christian Harder
Matt Artz
Keith Mann

Business

Catherine Ortiz
Jon Carter
Jason Childs

For information on Esri Press books and resources, visit our website at esri.com/en-us/esri-press.